How to Start and Operate an Electrical Contracting Business

How to Start and Operate an Electrical Contracting Business

C. L. Ray, Jr.

McGraw-Hill Book Company

New York   St. Louis   San Francisco   Auckland
Bogotá   Hamburg   London   Madrid   Mexico
Milan   Montreal   New Delhi   Panama
Paris   São Paulo   Singapore
Sydney   Tokyo   Toronto

Library of Congress Cataloging-in-Publication Data

Ray, Charles L.
  How to start and operate an electrical contracting business /
Charles L. Ray, Jr.
      p.   cm.
  Includes index.
  ISBN 0-07-051243-4 :
  1. Electric contracting—Management.   I. Title
TK441.R39 1989
621.319′2′068—dc19   88-7277
                                                                      CIP

1234567890   DOC/DOC   89321098

ISBN 0-07-051243-4

*The editors for this book were Harold B. Crawford and Georgia
Kornbluth, the designer was Naomi Auerbach, and the production
supervisor was Richard A. Ausburn. This book was set in Century
Schoolbook. It was composed by the McGraw-Hill Book Company
Professional & Reference Division composition unit.*

Printed and bound by R. R. Donnelley & Sons Company.

*For more information about other McGraw-Hill materials,
call 1-800-2-MCGRAW in the United States. In other
countries, call your nearest McGraw-Hill office.*

DEDICATION

To my wife Jayne; my parents, Charles Ray, Sr., and Ollie F. Ray; my sons, Charles Ray III and Christopher Ray; my sister, Hershal Lee; and my brothers, Larry Wood Ray, Danny Lyman Ray, and Ricky L. Ray. All helped me make special sacrifices. Thanks to God for this privilege, and may his richest blessings be with all of us.

Contents

Preface

Do you want to be successful in your electrical contracting business? If so, this book is for you!

Many electrical contractors started out in the field as skilled craft workers, working with their tools. Some were struck with the idea of going into their own business, doing all the work themselves, under cutting prices, and making *big* money. And, why not! This *is* America— the greatest country in the world—with the best free-enterprise system ever! This type of thinking about going into business is destined to fail. You must first have a success plan. You must be mentally tough and technically knowledgeable about what you want, where you want to go, and how you plan to get there. You must know where you are relative to where you're going. Remember: if you don't know where you're going, any road will take you there.

Just as in any worthwhile field of endeavor, the electrical contracting business has qualifications that must be met. At least 2 years of college, or the equivalent, is desirable. The training curricula should include electrical construction, trade mathematics and theory, drafting, and general business courses. Even if you complete the necessary schooling, on-the-job training, or apprenticeship, however, there is no guarantee that you will pass the electrician's test, the master's exam, or both. Commitment, dedication, and continued study are usually required to surmount these hurdles. Should you be unable to pass these exams, you may want to consider entering into a partnership with someone who can pass them.

You should have a reasonably good understanding of how to manage people, and how to get along well with them. Everything that will ever be accomplished on this planet will be done by, through, and with people; so, all of us must learn what we can about dealing with each other effectively.

Every entrepreneur entering the electrical contracting business should realize that records show that it takes approximately 10 years to become really viable. When you check out the profit margins in this industry, you will find that the national average is about 4 percent—

1.9 percent after taxes. However, if you don't have to borrow too much money, if you take advantage of all discounts, and if you conduct an efficient production operation, profits can run as high as 32 percent.

The electrical contracting business is very competitive. Performance standards must be excellent in all categories. When you have set your desire levels, disciplinary standards, and ultimate goals solidly in front of you, when you are confident in what you are doing, and when you are willing to take reasonable risks, you are on your way to success.

Businesspeople must organize their thinking and examine all aspects of their situation with meticulous objectivity. You may accomplish this by asking yourself a series of questions, the most important of which are

1. What is the situation or proposition under consideration?

2. What is the risk? What is the cost? What stands to be gained or lost?

3. Are there are any precedents? Can they be considered valid and applicable in this instance?

4. What do your competitors, customers, and other associates stand to gain or lose?

5. What are the known difficulties and obstacles? Consider in detail how they can be overcome.

6. Are any other problems likely to arise? If so, is risk capital available? How can you cope with these difficulties?

7. Are all the facts known? Are there additional, hidden pitfalls?

8. What is the turnaround time needed to accomplish the objectives or goals in question, should you decide to proceed?

9. Does the company stand to gain more by devoting equal time and effort to something else?

10. Are the personnel who are responsible for handling the situation fully competent?

Once you have the answers to these questions, you must weigh them to determine whether the undertaking is feasible. If the scales tip heavily in one direction or another, the decision is easy to make. If, on the other hand, the plus and minus factors tend to balance, then you must use your best judgment and decide.

C. L. Ray, Jr.

Acknowledgments

Over the years, hundreds of people in and around the construction industry have contributed to the writing of this book. I would like to extend my warmest thanks both to those mentioned in the book and to those whose contributions are reflected indirectly, as well as to the following people and organizations: Forrest S. Williams; E. J. Dunning; Rogers Johns, Sr.; Vincent Oliphant, Sr.; Peter Slovensky; Aubrey S. Slovensky; John A. Slovensky; F. W. Finney; Melinda Powers; Willia Poe; R. M. Buchanan; Loretta Mbah; Elizabeth Miller; R. W. Bowers; Wallace M. Grubb; Alfred Silverstein; Dr. Noel C. Taylor; Charles Reid and Joe Cullum of General Electric Supply Company; Cynthia Zenj-Ra; L. Wilson York, president of the First State Bank of Danville, Virginia; W. W. Emmerson; Margarite B. Poindexter; Dr. Richard L. Chubb; Greta Evans; Small Business Administration; National Electrical Contractors Association; Watson Electrical Co.

Special thanks go to A. L. Holland, Jr., who has unselfishly shared so many fruitful ideas with me throughout my contracting career, as well as to Howard L. St. Clair; Ronald J. Jordan; Ronald Neighbors; J. E. Barnes, Tony Vasfi; Zubaida Vasfi; J. H. Clements; M & T Electrical Enterprises; Luther A. Brown, Sr.; Moore & Campbell, CPAs; Tom Brown, Sr.; Keith Burns; Frank Aceves; and Herbert Mitchell. My special thanks also to Yvonne Dubose, Maxine Crowder Sprague, and Karen Blank for being patient and supportive with me in this effort. John C. Randall & Associates, Inc., has supplied valuable information on marketing. (For more specific information on marketing, write to John C. Randall & Associates, Inc., P.O. Box 15127, Richmond, VA 23227.)

I want also to extend special congratulations to two former employees: Bobby Terry, who now owns his own electrical contracting business, and Larry Woolard, one of the very few professional high-voltage and cable-splicing experts on the East Coast.

How to Start and Operate an Electrical Contracting Business

How to Succeed with People

Understanding Human Nature

If you wish to become a more skillful manager of people, you must first improve your human-relations skills. It is extremely important to understand people and their nature. You are on your way when you (1) begin to discover why people act the way they do, and (2) begin to understand why they respond as they do in various situations.

The art of understanding human nature is rooted in an acceptance of people as they are, without trying to impose your values on them. Inherent in human nature is a strong interest in self. The power of this trait is demonstrated when an individual receives greater satisfaction from the thought of giving a gift than from an appreciation of the value of the gift to the recipient. Once you become aware that people are more interested in themselves than they are in you, you are building the necessary foundation for dealing effectively with them.

The Art of Conversation

People enjoy having others speak well of them. When you give people direct compliments, you will most likely be rewarded with warm responsiveness in return. Your action and the receiver's response are customary tendencies of human nature. Anything else would be in contradiction to that tendency.

Good communication skills are also aided by references outwardly directed. Try using "you," "yours," and "yourself," instead of "I," "me," and "mine." This approach may be difficult for you in the beginning, but your ability to use it will improve over time. If you become more adept at encouraging people to talk about themselves, you will, in turn, grow in their esteem. Genuine inquiries about people's well-being create an atmosphere of harmony and goodwill.

Every human being wants to feel like "somebody." People do not enjoy being belittled or ignored. They want to be recognized, respected, and appreciated for their self-worth. However badly people may feel at any given moment, their morale can be lifted by cheery words of confidence. Generally, the more important and worthwhile you make people feel, the more favorably inclined they will be toward you. Give thought to your responses in conversation, so as not to seem disinterested. Refrain from giving anyone the brushoff, whatever his or her station in life. If someone has made an effort to see you, you should personally acknowledge that person's presence if at all possible. If not, direct a capable assistant to fill in for you.

Develop a frame of mind that allows you to agree. The art of openly agreeing with others can be the greatest human-relations asset you acquire during your lifetime. Don't be afraid to let people know when you agree with them. Say clearly, "You are right, and I agree with you." When you find yourself in disagreement, refrain from admitting it unless it cannot be avoided. Rather, try to present your different views with as much diplomacy as possible. At all times, avoid heated discussions which could lead to hot arguments. When you are wrong, admit it. Do not hesitate to admit your errors. Many people will use a cover-up or tell an outright lie. Do not do this.

A prime human-relations error is arguing. Do not argue. You will certainly not win friends or influence others if you insist upon arguing. Leave potential troublemakers and impulsive fighters alone. They will soon recognize the folly of their actions. If you keep yourself in harmony with others, you will find that you are liked by many people.

Good listening habits can help you become a better conversationalist and a more aware individual. Here are some valuable ways to achieve better listening habits:

1. Look directly at the person speaking. Allow no distraction to break your concentration.

2. Hold your body at an attentive, slightly forward angle to guard against becoming too relaxed.

3. Ask pertinent questions. This is an obvious indication that you are listening.

4. Do not interrupt, and do not change the subject until the speaker has finished.

Influencing Others

One key to influencing people is to first find out what they want. All of us like different things, and we react in response to different stimuli. We cannot assume that other people's interests will always be in tune with ours. We influence people best when we make a real effort to find out what they are seeking in life, what they really like, and where they want to go in their pursuit of health and happiness. After skillfully acquiring this information from people, you must respond positively to them by saying what they want to hear. You have to convince them (and this may well be the hardest part) that they can get what they want by doing what you want them to do. Herein lies the major component of the art of influencing people.

Here's an example: You are an employer in need of a truck driver. You know that some of your competitors are looking at the same driver you've been considering, and you must find out, from the viewpoint of the potential employee, what is most appealing in this position and in the company. Suppose that you find out that job advancement and job security are most important. You must be entirely convincing in proving to the potential employee how these needs can be satisfied best with your company, rather than with any of your competitors. By putting yourself in the other person's shoes, you can better sense real needs. If you honestly look at the person's concerns, your sincerity will be reinforced, thus giving you an advantage over your competitors with this potential employee.

In your efforts to convince others, be careful to say things that will benefit them and not yourself only. To do otherwise will arouse skepticism. If you cannot entirely avoid self-serving statements, temper the skepticism of others by quoting a third party. That is, let someone else make the statement for you even when that person is not present. For instance: You are asked about the quality and durability of a wristwatch. Rather than give a direct opinion, mention someone you know who bought a wristwatch like that several years ago which is still in good condition. You are not responding personally, and therefore, your credibility is not on the line. People are not quick to doubt a statement indirectly made.

Getting people to say "yes" demands more than pure luck. You must convince them of the importance of getting a particular job done. You have to drive home the advantages, and show how the job will best benefit them. Refrain from giving the reasons why the job will be of benefit to you. Pose "yes" questions by setting up a positive frame of reference from the start, through pointed questioning. Remember to nod your head affirmatively as you ask questions such as "You surely would like a happy family, wouldn't you?" and "You surely want the best pay for your labor, don't you?" The "yes" question, as you can see, is one which in all probability can only be answered with a "yes" answer. People like to be given choices. Be certain to give them choices which must be responded to with a "yes" answer.

Here is an example: You want to visit someone about a business deal. Your question should be constructed this way: "Which time will be better for my visit, 2 p.m. today or 3 p.m. tomorrow?" You cannot expect an affirmative answer on every question, but with this method, you are certain to get a positive response most of the time. Expecting people to say "yes" is a show of confidence. Most people respect and respond to proven leadership. Knowing this, you can gather up the wherewithal to provide that expected leadership. People tend, usually, to react in kind to the positive or negative behavior of others. With this in mind, be advised that what you put out will most likely be what you will get back. A first encounter can set the pace for what's to follow. At the initial point of eye contact, send out your most engaging smile. More than likely, your smile will be returned. At this point the mood is being set, the pace has begun. If you are smart, you will realize that this important gesture establishes a mood that will benefit you while also benefitting people you are trying to impress. The all-important smile should always precede the beginning of the conversation.

Sincere Praise

No one lives by bread alone. Do you remember the last time you received a kind word for a deed well done? Do you remember how good you felt? Kind words that have been earned go a long way. As you consciously observe people's deeds for the purpose of finding opportunities to extend praise, you'll become more and more aware of the effectiveness of kind words, and you will act accordingly. Your praise will be genuine and sincere. However, if there are instances in which you cannot praise genuinely and sincerely, it is better to say nothing.

In addition, remember to praise the act performed rather than the individual. You can thus avoid any hint of favoritism. And praising the act rather than the doer also encourages favorable behavior in the future. Some examples:

> *Say*: "Bill, your work this past week has been excellent!"
> *Don't say*: "Bill, you're a good guy."

or

> *Say*: "Jenny, your artwork is simply beautiful."
> *Don't say*: "Jenny, you sure do work hard."

Try using the "happiness formula": That is, get into the habit of giving people a word of praise. They will feel good and so will you. Remember that the real joy in life comes from giving, not from receiving.

Positive Criticism

Another useful thing to remember is that nobody appreciates being negatively criticized. You should never criticize someone to inflate your own power quotient. If criticism is to be of any value at all, it should be of the positive variety only. The following suggestions should prove helpful:

1. Criticism should be given only in privacy and between the persons involved.
2. An atmosphere of warmth and civility should be set before the criticizing takes place.
3. Only the act should be criticized, not the individual.
4. After offering helpful criticism, suggest or request the correct or a more effective course of action.
5. Never demand cooperation; ask for it. You will have greater success with this approach.
6. Criticize only one act, and in a noncondescending manner. Try to generate an air of friendliness at the end of the meeting, if at all possible.

These suggestions should help you to gain the cooperation of the individual whose work was criticized.

Expressing Appreciation

Another facet to an understanding of human nature involves the art of skillfully thanking people. It is not enough to feel grateful toward someone. You must seek ways to show your gratitude. People tend to give more and do more when they feel genuinely appreciated. Why? Because it's human nature to do so. When you once fail to express gratitude to a person who has done you a favor, you rarely get the opportunity to express it again. However, remember that most people can sense sincerity—and lack of sincerity. If your appreciation is not sincere, you had better remain quiet.

Speak audibly and distinctively when you thank someone, so that each word will be heard and understood. Use the person's name; this adds personalization. Practice thanking people properly and often. The habit of expressing appreciation generates goodwill and is generally beneficial for both you and the people you thank.

Making a Good Impression

It is important to acquire the art of skillfully making a good impression. Most people would answer "yes" to the question "Do you want to be admired, looked up to, praised?" However, to merit this recognition, we must continually behave well, so as to be worthy of praise. All of us, in various ways but primarily through our behavior, control other people's opinions of us. If we are wise, we behave in such a way as to favorably impress people. For example, suppose you are asked your occupation and you answer, "Well, I'm just another keyboard puncher." This statement would never impress anyone. Your answer should have been: "I am very fortunate to be working for XYZ Company. They use skilled people to test their keyboards."

At all times, try to exude confidence in what you do and show enthusiasm in what you say. It's contagious and, most of all, it works. Enthusiasm is a most valuable tool when selling yourself or your product.

Public Speaking

The art of public speaking is extremely important. Here are some key suggestions that can make addressing a group easier:

1. Be acutely tuned in to what your audience wants to hear.
2. Get to the point quickly with clear and concise statements.
3. Lean forward as you speak, and look directly at the people in your audience.
4. Remember that what's of interest to the audience is what counts, not what's of interest to you.
5. Try to give a warm, intimate talk when possible, rather than a stiff, formal speech.

Summary

Finally, here are a few thoughts to keep in mind:

1. Knowledge is power. But neither power nor knowledge will do anyone any good unless it's used—advantageously.
2. You are to be commended for improving your knowledge of human relations.
3. Practice, practice, practice.
4. Love people and use money. Never love money and use people.

Getting Started:
Financial Planning

Initial Financial Planning

For the sake of both survival and financial success, management must constantly review a company's financial situation. Adjustments should be made to assure skillful planning combined with a pattern of controlled growth, not to exceed 5 to 7 times net worth. The growth pattern should build operating capital while maintaining a minimum of 5 percent profit on gross sales—enabling the company to expand on its own money, and not on the lender's.

Planning involves two stages: short-range planning, which is no more than a monthly budget, and long-range planning, which involves 5 to 10 years. Even the best plans, however, cannot be rigidly implemented. In company planning, as in traveling, a detour is sometimes necessary, but ultimately, you must return to the main road.

To begin company planning, you will need a plan-of-management chart which clearly shows the responsibilities of each person in your company (see Chapter 3). You will also need both a clear-cut employee policy manual with precise job descriptions and an employee handbook for each person in your organization (see Chapter 4). Learn all you can about human needs—which are basically the same for all people, whether they occupy superior or subordinate positions in the company hierarchy—and about how you can assist individuals to fulfill their own needs, so that the company will get the benefit of each employee's most productive work.

Financial and Legal Structuring

Once the initial financial plan is made, you need to deal with the legal aspects of setting up and running a business. It is important to have a construction attorney you can trust, preferably one who is knowledgeable about the legal aspects of the electrical contracting industry. In addition, you will need to consult with a good certified public accountant (CPA) who has experience in the construction industry. The firm Moore & Campbell is one of the best CPA firms on the East Coast for construction firms, and particularly for electrical contractors. Drop a line to Jay Moore; tell him I told you to.[1]

The basic structure you set up with the aid of the attorney and the CPA will cost no more for a 2- to 5-person operation than it will for a company of 100 or more employees. Should the company grow to 100 or even 500 employees, you will be able to utilize the same basic structure with only slight modifications.

In case you are about to go into business for yourself for the first time, a word of caution: The cheapest money you'll ever spend will be for some of the most expensive services you will use—those of the accountant and the attorney. If you do not use them, you will pay the IRS and others twice what you would have paid for these professionals.

Initial Operating Funds

Your next important step is getting money—from your own savings or from other sources—to start up and operate your business. You will need funds to cover a minimum of 252 workdays (12 months) as if you will bring in no money at all during this period, or as if you will have to wait for your money. And quite frankly, you should ideally have sufficient resources to last for 2 to 3 years while you build the business. But if you do have these resources, it is best not to let your customers know it. Here's an old saying that is true: "The squeaky wheel gets the grease." Demand your money in 30 days or less.

Financial Management

Successful businesses plan and schedule effectively. To do this, you must have very good discipline. The construction industry has an unfortunate, but well-deserved, reputation as a high-risk, high-failure industry. It is one of the largest, most competitive and most difficult-to-control industries. Considering the size of the industry, the total need for good financial planning is tremendous. When the industry's size and high-risk performance are combined with closely held ownership, the need for expert financial management becomes even more apparent.

Bonding

Contract performance guarantees by bonding companies require that you provide financial statements in addition to a lot of other data to

---

[1]Moore & Campbell, P.O. Box 7458, Roanoke, VA 24019.

assist the surety in its monitoring of operations. To learn more about this matter, send for a copy of *Bonding Manual*.[2] For alternative bonding sources relative to government contracts, write to the author.[3]

Tax Planning

In a closely held construction company, tax planning is another major consideration, and your CPA can best guide you in this area.

Causes of Business Failure

Stop and think for a minute about the businesses you know of that have failed. Why did they fail? It is not so surprising to learn that, according to government statistics, the principal causes of business failure are poor management and undercapitalization. The extensive comments on this subject in this book are an attempt to cover the most important areas of financial management. Some areas will be covered in depth; in others, we will deal with the high points only. Though this book deals specifically with the electrical construction industry, the guidelines provided here can also benefit the architect, the engineer, the CPA, or even the supplier. In addition, people working in these allied fields must understand the nature of the electrical contracting industry they are serving if they are to succeed in their own endeavors. The primary objective of everything discussed here is to develop a base upon which financial planning can be built.

Financial Reports

An essential ingredient of financial planning is a management information system which (1) develops key data necessary to make decisions, (2) summarizes the data in an understandable form, and (3) delivers the data to the appropriate person on a timely basis. Such a system should include three types of reports: (1) what has happened, (2) what is forecasted to happen in the near future, and (3) long-term planning and corrective action.

The first type of basic report relates what has happened in the latest reporting period. This cost report enables managment to converse with the production supervision and estimators for the purpose of comparing the estimated with the actual cost.

The second type of report reflects what is forecast to happen in the near future. However, there is no magic crystal ball which will predict the future; it has been said many times, in fact, that the only way to judge the future is by the past. This simply means that the past his-

---

[2]Write to Tom Brown & Company, Inc., P.O. Box 19293, Washington, DC 20036

[3]C.L. Ray, Jr., 6949 Connie Drive, N.W., Roanoke, VA 24019

torical performance of a company can reasonably be used as a guide to indicate future performance.

The third type of report includes long-term planning and corrective action. Long-term planning is described in Chapter 8. Corrective action may be indicated on the basis of the above reports, when actual performance and costs are compared with estimated cost.

Of course, each company must decide which reports are essential for use by each level of the management team. Reports must be geared to their readers. Top management's concerns may not be the same as those of the project manager or the field supervisor.

Individual reports may reflect current status of operations for the month or the year to date, or may analyze data. A basic report that is very meaningful to management is the monthly job-cost analysis. This is a job-by-job breakdown of actual costs, analyzed by category and compared with originally estimated cost-to-date figures. The difference between actual and estimated cost should be shown for each item. If resources permit, estimates can be updated periodically.

Financial reports serve to show the financial condition of a company as of a given date. In their various forms, these basic reports tell where a company is, where it has been, and where it is going. Financial reporting is an accounting function consisting, basically, of a balance sheet, a statement of retained earnings, and a statement of changes in financial position.

Accounting is not a science; it is an art—an assumption, an opinion. All key individuals on your company's management team must understand accounting reports if they are to serve their purpose. Again, failure of management to understand financial planning leads the company down the road to nowhere. Intuition is a very poor substitute for sound, intelligent evaluation.

Planning for Profits

Selecting the proper method of financial reporting is critical. A major problem is recognizing when the profits of a firm will be realized. There are four ways to recognize profits: (1) cash method, (2) accrual method, (3) percentage-of-completion method, and (4) completed-contract method.

As always, the secret of success in business is: Plan! Plan! Plan! If you really want to know where your business stands financially, ask yourself if you could survive for 120 days if your cashflow were suddenly cut off.

Your main objective should be net profit. But making money is one thing; holding on to it is another thing entirely. It may surprise you to know that many contractors handling several million dollars worth of business cannot tell you whether they made a profit or not until the end of the year. I have heard contractors say, "I've got to see my accountant to find out whether I made any money." The fact is, however, that if contractors are reasonably well attuned to their job costs and weekly reports, they know whether or not they are making money. It would make better business sense to hear a contractor say, "I've got to see my accountant to find out how much money I made (or lost)."

Financial Ratios

One discovery I have made, after working as a consultant and an electrical contractor for 25 years, is that the smaller the business, the more likely it is that management is not monitoring ratios and their meanings. And the more likely it is that management is not marking up prices to cover overhead and make a profit. Of course, in a small business it is also easier to check on whether you have made a profit on your overall operation. But this is not enough! You need to know why you did or did not make a profit on a certain job, and you need to know what phase of the job caused you to make or not make the profit. You should be finely attuned to your business, and you should control its operations by using financial ratios, not intuition.

The Five Critical Areas of Financial Management

In financial management, there are five areas that management must stay on top of:

1. *Measuring financial performance.*   These measurements are made through the use of ratios and balance sheets.
2. *Managing profits.*   Normally used for working capital or put in certificates of deposit for bonding liquidity. Also, unless you understand your company's breakeven point as it relates to volume and profit, you probably won't make a profit.
3. *Managing assets.*   The success of this area is usually based on your return on investment (ROI).
4. *Budget and control.*   This is a form of short-range planning; the results are measured against day-to-day operation through reports.
5. *Managing cash.*   This is a form of cashflow forecasting. Funds must be available to cover day-to-day operations. An acute shortage in this area can be compared to an acute shortage of blood in the human body. Circulation is cut off, and everything else stops working. Management must have an intelligent understanding of

cashflow, and cashflow analyses should be done on each job. This is really the only way your banker will be convinced that you know what you are doing—and your banker's opinion can count for a lot.

These five areas make up the financial-planning aspect of your business. For a discussion of estimating as it relates to financial management, see Chapter 5.

3

Standard Organization and Procedures for the Electrical Contracting Business

The Plan-of-Management Chart for Sample Electric Company

When a plan-of-management chart is prepared for any business enterprise (especially where this procedure is a new development in the company), the need arises for personnel to study the chart in order to learn about the structure of the organization, about the chart itself, and about the benefits to be gained from a sound understanding of the principles underlying the authority and responsibility that flow from and to management—the principles that achieve a prime function within the company as a whole. Understanding the plan-of-management chart will give personnel an accurate concept of the entire plan of management and the structure and limits of authority.

The plan-of-management chart, or functional organization chart, an example of which is shown in Figure 3.1, is a graphic portrayal of the relationship between the various functions within the entire organization.

- The chart is management's pattern for allocating authority and responsibility within the organization. It provides the necessary conduits, points of origin, and terminations for the flow of all management directives and controls between key personnel and management.

The plan-of-management chart for Sample Electric Company (Figure 3.1) contains titles, which represent the functions responsible directly

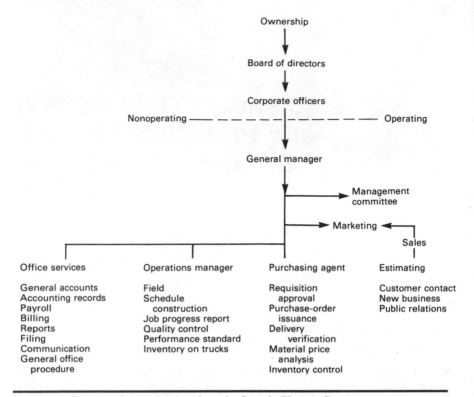

Figure 3.1   Functional organization chart for Sample Electric Company.

to management, and channels, which represent the lines of authority and responsibility. It is important to understand that a title represents a function—not the individual who carries out the function.

Flow of Authority

At the top of the chart is the nonoperating function: ownership. Ownership selects and appoints the general manager, who has full and complete authority over all operating functions of the company.

Selected by and responsible only to the general manager are the heads of the prime functions. To each such head is delegated the authority and responsibility commensurate with the position. Below each activity head are the direct functions (or departments), each of which is the direct responsibility of one person, who is selected and appointed by the function head.

Line of Responsibility

There is only one line of authority and responsibility in the organization; one person may fulfill more than one function, but one function is responsible to no more than one other function. The reason for this is that *one person cannot do two jobs at the same time.*

The Importance of Delegation

Sound organization requires that delegation of responsibility be basically complete and that each position in the organization be given the maximum possible autonomy. The organization which fails to delegate limits its effectiveness to the talents and energies of a few individuals. In contrast, the organization which delegates effectively is in a position to completely utilize the energies of every member of the organization. For maximum effectiveness, every individual must be free to have failures as well as successes, and it is essential that each individual be judged by results over a period of time, and *not by a single event.*

Delegation of responsibility requires equal delegation of authority. Authority includes the right to direct, coordinate, and determine, but does not imply autocracy. A superior's invasion of the area of authority of a subordinate has the effect of temporarily relieving the subordinate of responsibility, but without a definite and continuing assumption of that responsibility by the superior. Such an invasion causes confusion and loss of effectiveness.

Synthesis of the Organization Chart

The list below will serve as a guide in reading the chart, and will help in acquiring an understanding of the relative positions of the various functions.

1. Down the channels from the top of the chart flow directions, policies, duties, responsibilities, and controls which management exercises.

2. Up the same channels flow the reports, records, and schedules which are necessary to keep management informed. These communicate the results of operation and ensure the implementation of directives that have passed down the line.

3. Orders pass directly through the function which the channels connect. Each function sorts out the orders and acts upon them, making decisions and accepting the responsibilities involved or passing them up or down the channels, as the case may require.

4. There is no relationship of authority between jobs on the same level. There is, however, a nonauthoritarian relationship between functions in the organization, which is not shown on the chart. These functions have no authority over one another; the relation-

ships between functions must be maintained through cooperation and mutual assistance.

Organization and Ownership of Sample Electric Company

The purpose of this standard procedure is to establish the relationship between ownership, management, and the operating employees of Sample Electric Company.

Ownership

Authority and Status.   The authority of ownership is and must be absolute. There is no possible recourse to a higher authority than the stockholders of this corporation.

The function and status of ownership are continuous, even though the individuals involved may change from time to time. There will always be capital stock, indicating the presence of ownership.

Ownership is interested in a reasonable return on its investment, adherence to established broad general policies, enhancement of the corporate image, the best possible pay and working conditions for employees, conservation of assets, replacement of assets, expansion of the business, and above all, maximum service and satisfaction to the customers of the corporation.

Ownership has the responsibility for determining and providing those general policies which will permit the organization of Sample Electric Company to operate within the limits prescribed by federal and state laws.

Duties.   In order to achieve the above aims, ownership, composed of the stockholders, selects a board of directors. The board of directors elects a president, vice president, secretary, and treasurer. The president then has the authority to select and, with approval from the board of directors, appoint a general manager. The general manager shall, with recognition of the major function of the business which exists by virtue of the nature of the business, select and, with the approval of the president, appoint a key person to each of the major functions of the operation.

The conditions of appointment of the general manager shall be with an agreement that a predetermined reasonable profit can be generated by use of the available working capital, land, leases, buildings, equipment, and inventories of Sample Electric Company.

The general manager, upon appointment, shall have full authority and responsibility to do whatever will earn a profit, so long as it is ethical and in the best interests of the company ownership. Ownership shall not abrogate the responsibility of the general manager by dictating minor policies.

Ownership must be unanimous in deciding and acting upon all issues and performances for the good of Sample Electric Company. Ownership must take proper action to secure an adequate return on the capital invested in Sample Electric Company.

Ownership must take proper action to preserve the reputation and prestige of Sample Electric Company.

Delegation of Authority and Responsibility. In delegating authority and responsibility, ownership must conform to federal and state laws. Ownership can delegate its authority as member of the board of directors, as well as at all other meetings attended by ownership, if such delegation is legally permissible.

General Manager

The purpose of this standard procedure is to describe the activity area, authority, responsibilities, and duties of the general manager of Sample Electric Company.

Authority. The general manager is appointed by and responsible to ownership for performance of the assigned duties and for fulfillment of the responsibilities of the position.

Responsibilities. The general manager is completely responsible to ownership for the activity areas pertaining to the position of general manager, and for the coordination and successful operation of the company.

The general manager is responsible for the performance of subordinates, and for their implementation of and compliance with the operating policies established for the company. The general manager is responsible for the conservation of company assets and for the generation and maintenance of ample working capital.

Duties. The duties of the general manager include, but are not limited to, the following:

General

1. Direct the planning and scheduling of all operations through departments and their key people, which together make possible full control of all operations.
2. Enforce adherence to all plans, policies, schedules, and directives.
3. Exercise control over the number of people and the amount of expense required to support a planned level of activity and profit.
4. Hold regularly scheduled management committee meetings, including all department heads, for the purpose of coordinating and using to best advantage the combined thinking abilities and experience of those people.

Personnel

1. Establish and maintain a strong policy of hiring and keeping a high caliber of personnel, from key functions down through the lesser jobs in the company.
2. Adopt sound, modern, and competitive personnel policies, programs, and procedures, including a formal on-the-job training program.
3. Set salary and wage levels, and make periodic adjustments as conditions warrant.
4. Review with the department heads, as required, the selection, employment, promotions, discipline, and dismissal of employees.

Marketing

1. Study the business potential of the company relative to the total available market.
2. Define the market goals and establish sales quoted by logical breakdowns and customer categories.
3. Set up a planned program of sales development, including personal contact.

Operations

1. Schedule, budget, purchase, and have available any equipment or facilities necessary for the profitable operation of the business.
2. Set up and maintain, as required, a preventive-maintenance program.
3. Plan, forecast, and control cashflow to ensure its availability for purchase of equipment or services needed, with a minimum of borrowed funds.
4. Establish and maintain a program of good housekeeping and safety.

Control

1. Plan and control the finances of the company so as to avoid, where possible, debt financing, and so as to allow for growth and expansion.
2. Control cashflow in such a manner as to assure the regular discounting of bills.
3. Schedule operations on a planned program to produce annual earnings after income taxes at an absolute dollar.
4. Set up budgeting procedures that will control all operations so that a predetermined profit will be recognized.

Finally. It is self-evident that not all the particulars of this all-embracing function can be reduced to writing. However, when the initiative, drive, task, energy, vision, and other necessary qualifications of the general manager are coordinated and directed into this one channel, the top function of general manager becomes the most important single factor contributing to the overall growth, stability, profit, and success of the company.

Management Committee

In a company large enough to be divided into functions and departments, communciation and coordination can be very difficult. The problem increases in direct proportion to the continued growth of the company. The best device for dealing with this problem is a management committee. It is the purpose of this section to describe how a management committee operates.

The management committee is directly responsible to the general manager, who serves as its chairperson. The membership consists of the heads of all departments. The committee is advisory only. Final decisions are made by the general manager, and the rulings of the general manager are binding on all employees, including members of the management committee, regardless of their personal views.

The management committee meets weekly at a time and place designated by the chairperson. The time and place of the meeting are on a standing basis, with no change from week to week; thus members are on notice to appear at the meeting every week, and no weekly notification is necessary.

The meeting should be short: 30 minutes. The session should start and stop exactly on time. If the topic under discussion has not been concluded, the meeting is to adjourn anyway and the matter be resolved at the next meeting.

Topics to be considered are those problems crossing departmental lines, or any difficulties, problems, or other topics the discussion of which should improve company profits. Responsibility for solution of a departmental problem always remains with the department head. All members of the management committee will contribute their knowledge and ideas to the resolution of each problem. However, this in no way relieves the department head from ultimate responsibility which cannot be transferred.

Above all, it must be understood that the general manager cannot solve all problems. Modern business is too complex for anyone to be capable of doing this. The head of each department is expected to specialize in his or her own area of activity and to be the outstanding authority in the company on that area of specialization. The chairperson (or general manager) will weigh the studied opinions of department heads on companywide problems, and select the line of action which appears to offer the greatest possible chance for success.

Sales Manager

The purpose of this standard procedure is to describe the activity area, authority, responsibilities, and duties of the sales manager of Sample Electric Company.

The sales manager is appointed by and reports to the general manager. The sales manager and the general manager work hand in hand to plan the sales strategy so as to bring in more and more business volume.

The sales manager is the business promoter of Sample Electric Company. As compared with the situation of sales managers 20 years ago, present-day sales managers face tough and cutthroat competition. Thus, sales managers today have to make a far greater and more concentrated effort to bring in the business. To survive, they have to add to the existing business—and so they must survey new areas and territories. At the same time, they cannot afford to lose existing clients through complacency. They must display their products attractively and completely in order to demonstrate their ability to meet clients' needs and help them decide to use the services of Sample Electric Company.

Sales managers must have a thorough knowledge of all aspects of their own businesses and also have some idea of their competitors' businesses. They must have a very good understanding of the costing system. They must take guidance from past sales analyses and must be able to forecast future trends.

A sales manager must dress neatly and present a smart appearance in order to be able to use personal magnetism to influence customers. Customers must get the feeling that they are dealing with a progressive company's salesperson in whom they can place full confidence about time and cost matters. A sloppy appearance makes customers feel that they are dealing with an unreliable and deceitful person.

Successful salespeople, as a rule, walk and talk with an air of confidence which inspires not only themselves but also their customers. They are well-behaved and have commendable manners. They are forceful and enthusiastic, and spread their enthusiasm wherever they go. Good salespeople always close the deal in their own minds before approaching a customer.

Authority. The sales manager has ample delegated and implied authority to comply with the duties and responsibilities described below.

This standard procedure comes into full force and effect when signed by the general manager and remains in full force and effect until modified by the general manager.

Responsibilities

1. Bring in an increasing volume of business every year and meet a target percentage every year.
2. Obtain reasonable prices commensurate with competition.
3. Pay special attention to high-profit items.
4. Add new items and new customers, and take special care of them.
5. Coordinate sales with service capabilities to meet delivery deadlines.
6. Create an impression on the customer that emphasizes the reliability, progressiveness, and high capabilities of Sample Electric Company.

Duties

1. Conduct sales promotions and surveys.
2. Provide for sales diversification.
3. Promise and stick to delivery schedules.
4. Give prompt and reliable replies to customers.
5. Coordinate with the general manager on matters of advertising, costing, and forecasting the sales.
6. Never neglect a customer's complaint even when the customer is at fault.
7. Compile and analyze statistics on customers, type of products, high-profit items, and developing territories.

Office Manager

The purpose of this standard procedure is to describe the activity area, authority, responsibilities, and duties of the office manager of Sample Electric Company.

The office manager is appointed by and reports to the general manager. The office manager keeps all the records pertaining to the control function, bookkeeping and accounting, liaison with auditors and public accountants, profit and expense control, payroll, accounts receivable and accounts payable, and invoice verification. The office manager also fills in the necessary forms to comply with state and federal regulations.

In addition, secretarial work, documentation and filing, incoming and outgoing mail, reception, inquiries, and telephone communication with the public are all taken care of by the office manager. Dealing with the public is an important aspect of the office manager's responsibility, since people from outside the company, whether they telephone or visit in person, form an impression about the company which is greatly affected by the treatment they receive. A cordial attitude does much to creates a favorable impression.

The office manager, being in direct touch with expenses, must bring excessive variances to the notice of the general manager. This function is very important in that it enables the general manager to take action to control excessive and unnecessary expenses before it is too late. All departments must get information on expenses from the office manager as soon as possible, in order to decide upon and apply the necessary controls under their jurisdiction.

The office manager must make a detailed list of miscellaneous expenses at least once a month. Though these expenses are generally small, they sometimes add up to a sizable amount which, if controlled, can result in net savings.

A separate record must be maintained for accounts receivable. Reminders for recovering the arrears must be mailed promptly at the appropriate time.

Payroll calculations must be correct and issued on time so as to avoid complaints, frustrations, and demoralization of employees.

The office manager must be aware of the basic fundamentals of good supervision and must have sufficient confidence and practice in the use of the same. Smart office managers, by using their own initiative and judgment, can effect tremendous savings. The office manager who believes that the duties of an office manager are merely clerical and routine is not truly functioning as an office manager.

Authority.   The office manager has full authority, implied or otherwise, based on the functional aspects of the job. The office manager must not tolerate any interference with the functions described here.

This standard procedure comes into full force and effect when signed by the general manager, and remains in full force and effect until modified by the general manager.

Responsibilities

1. Receiving customers cordially and giving them the information they desire.
2. Timely calculation and issue of payroll.

3. Detection of excessive expense on a timely basis, and bringing it to the notice of the people concerned.

4. Proper accounting of all money received and paid.

5. Proper filing system and documentation of important papers.

6. Prompt control of incoming and outgoing mail.

7. Gathering full information on and complying with all tax and government regulations.

Duties

1. Reception and inquiries.

2. Running the telephone switchboard.

3. Keeping ledgers and books.

4. Keeping records on accounts receivable and accounts payable.

5. Liaison with auditors and CPAs.

6. Performing all secretarial work.

7. Documentation.

8. Keeping payroll records.

9. Keeping inventory records.

10. Taking care of incoming and outgoing mail.

11. Compliance with state and federal regulations.

12. Taking care of state and federal tax matters and insurance.

Operations Manager

The purpose of this standard procedure is to describe the activity area, authority, responsibility, and duties of the operations manager (OM).

The operations manager covers all the contracts with the five Ms: men and women, material, methods, movements, and machinery. The operations manager plans those contracts; decides upon the personnel needed to complete the jobs within the allotted time limit; and coordinates the flow of materials, methods, movements, and machinery.

Operations managers must have a good psychological approach in order to deal with various contractors who have different temperaments, whims, and attitudes. They must also train their subordinates to deal effectively with fault-finding and "fuzzy" people (that is, people who are unclear about their needs and wants) . Without proper training, dealing with such people may upset subordinates and may result in work slowdowns, faulty work, or both.

The operations manager must provide on-the-job training so that assistants can carry on the work independently, with as little supervision as possible. This is absolutely necessary because the operations manager cannot be present at all the different work sites to solve the problems. However, the operations manager must visit the work sites at suitable intervals, both to get firsthand knowledge of the activities at the work sites and to establish contact with the contractors and find out their opinions about the job.

Operations managers in electrical contracting companies must have electricians' cards from the county and the city, and considerable practical experience is also necessary to enable them to face contractors' criticisms. These managers are also responsible for ensuring that valuable profits are not eaten away by *callbacks* and *defective work*. They must plan the work of subordinates by the hour and must set up a control system of feedback to measure performance. They must see to it that their people are fully equipped with the proper tools and equipment and are very punctual about reporting to the work site. At the same time, operations managers must be available for decisions in case they are called on the radio equipment or telephone. They must have a thorough knowledge of the costing system and of standard times for performing different types of work. They must set up suitable theoretical, practical, and oral tests to judge the desirability of promoting and giving increments to subordinates—for good work must be rewarded.

This standard procedure comes into full force and effect when signed by the general manager, and remains in full force and effect until modified by the general manager.

Authority.   The operations manager has full authority and jurisdiction for the operations of the corporation. This authority is vast, but the operations manager must use it with tact and diplomacy.

Responsibilities

1. Ensure the correctness and accuracy of the work.
2. See to it that subordinates have sufficient work at hand.
3. See to it that subordinates do a fair day's work.
4. Ensure that callbacks and corrections are minimal.
5. Show profits in operations.
6. Maintain sufficient inventory of common items.
7. Ensure that all tools and equipment are in good repair.
8. Keep a sufficient number of trained personnel on duty.

9. Plan work well in advance for personnel and equipment needs.

Duties

1. Inspect the materials required at appropriate times.
2. Plan the work of subordinates.
3. Conduct liaison with contractors.
4. Train personnel.
5. Maintain standards of discipline, order, and punctuality.
6. Assure the safety of personnel and equipment.
7. Establish work performance standards.
8. Comply with state, county, and federal regulations.
9. Plan his or her own work to allow time for visiting the different sites.
10. Make reports on the progress of the different jobs and discuss them with the general manager.

Purchasing Agent

The purpose of this standard procedure is to describe the authority, responsibility, functions, duties, and responsibilities of the purchasing agent.

Authority. The purchasing agent handles procurements—securing, when required and at minimum cost, the quantity and quality of materials, supplies, services, and equipment needed to operate the company. The purchasing agent is appointed by the general manager.

Responsibilities. The purchasing agent is completely responsible to the general manager. The responsibilities of the purchasing agent are as follows:

1. Locate supplies and secure quotations.
2. Work with the general manager on selection of supplies, and prepare purchase agreements.
3. Advise on specifications for purchased items.
4. Ensure that purchased items and services arrive as promised by vendors.
5. Analyze the need for purchased items, and secure items when required.

6. Maintain vendor catologs, quotation requests, purchase orders, material specifications, and the like.

7. Analyze market trends, purchased-item usage, buying, methods, and the like. Recommend improved purchasing practices when needed.

8. Maintain harmonious working relations with sales representatives from supply companies.

9. Strictly observe lead times in purchasing to guarantee delivery consistent with work-completion schedules.

10. Maintain predetermined inventory levels on a current basis.

11. Determine minimum-order and price-break quantities, and order accordingly.

12. Keep up inventory-control cards.

13. Assure that stocks will meet demands.

Duties

1. Purchase research.

2. Purchase material and equipment.

3. Purchase expediting.

4. Purchase records and files.

5. Inventory control.

Profit and Loss Control for the Very Small Contractor

Introduction

Most companies spend thousands of dollars each year on operating and general and administrative expenses. While some of these outlays are beyond control, a company must be able to rigidly control and, wherever possible, reduce those expenses which are within its control.

Profit and expense control involves formulation of anticipated expenses at each level of operating revenue, from 50 to 150 percent of normal operations in 10 percent increments. Once the level of revenue activity for a month has been determined, actual expenses may be compared with forecasted expenses to determine why variations have occurred and to reduce expense on controllable items.

Concept of Profit

The profit and expense control system places profit in the position of the first expense of business. Profit must not be regarded as a residue after expenses. If a company cannot assure a reasonable return on its services and investments each year, it cannot remain in business. Profit is a necessary reward of investment; without profit, no business will survive.

Profit and Expense Control Schedules

Profit and expense control schedules are prepared after a thorough review of the company's operating history. A scientific analysis of revenues and expenses is necessary before the schedules can be prepared. Expenses are reviewed realistically. All the established goals are attainable through reasonable actions and efforts by management to produce desired results. Once prepared, profit and expense control schedules become an effective guide to expense controls. No longer will expense levels be accepted without question or review. Necessary expenses will be expected, but unnecessary expenses and waste will not be tolerated.

Putting the Profit and Expense Control System to Work

On or before the 25th of each month, the general manager will estimate the level of operating revenue for the coming month. The bookkeeper will then enter the net operating revenue estimated on the monthly profit and expense control schedule (Figure 3.2).

The estimated level of activity is determined by matching the estimated net operating revenue with the net operating revenue listed on the control schedule that is closest to it, and then by interpolating to determine the level of each item of revenue and expense. See Figure 3.2.

A schedule of 1 percent interpolation factors appears in the column on the right of the control schedule. By dividing the operating revenue at 100 percent into the estimated operating revenue, we obtain the operating percentage at the estimated level of activity. We multiply this percentage by each available interpolation factor, and the result is the income or expense factor at the level of estimated activity.

Where no interpolation factor exists, see Figure 3.2 and proceed as follows:

1. Enter the activity level closest to the activity.

2. Subtract the next level below.

3. This equals 10 percent.

4. Divide the answer by 10.

	ACTIVITY	1	2	3	4	5	6	7	8	9	10	11	12	13
		50%	60%	70%	80%	90%	100%	110%	120%	130%	140%	150%	REMARKS	1%
300	SALES	12500	15000	175000	20000	22500	25000	27500	30000	32500	35000	37500		25000
	COST OF GOODS SOLD													
400	Purchases	5250	6300	7350	8400	9450	10500	11550	12600	13650	14700	15750		10500
403	Direct Labor	3375	4050	4725	5400	6075	6750	7425	8100	8775	9450	10125		6750
	TOTAL	8675	10350	12075	13800	15525	17250	18975	20700	22425	24150	25875		
	GROSS PROFIT	3875	4650	5425	6200	6975	7750	8525	9300	10075	10850	11625		
	OVERHEAD EXPENSES													
501	Salaries - Officers	1354	1354	1354	1354	1354	1354	1354	1354	1354	1354	1354		-
	- Office	675	675	675	675	675	675	675	675	675	675	675		-
405	Payroll Tax - Workers	200	240	280	320	360	400	440	480	520	560	600	5.9% FICA	400
408-9	Insurance - Workers	112	130	150	168	204	223	242	260	280	300	325	16.88 + 1.6 Ea.	223
410	Repairs - Auto	60	71	82	93	104	115	126	137	148	159	170		115
412	Gas and Oil	312	374	436	498	560	622	684	746	808	870	932		622
420	Permits	100	100	150	150	200	200	200	250	250	300	300		-
425	Bad Debts	50	60	70	80	90	100	110	120	130	140	150		100
503	Payroll Tax -Officers	120	120	120	120	120	120	120	120	120	120	120	5.9 FICA	-
505	Telephone & Telegraph	60	60	60	60	60	60	60	60	60	60	60		-
506	Office Supplies	57	67	77	87	97	107	117	127	137	147	157		107
507	Professional Services	115	115	115	115	115	115	115	115	115	115	115		-
508	Rent	325	325	325	325	325	325	325	325	325	325	325		-
509	Utilities	68	68	68	68	68	68	68	68	68	68	68		-
515	Depreciation	645	645	645	645	645	645	645	645	645	645	645		-
520	Dues & Subsciptions	35	35	35	35	35	35	35	35	35	35	35		-
522	Insurance - Officers	82	82	82	82	82	82	82	82	82	82	82	33.76 + 11.00 Ofr.	-
523	Postage	20	20	20	20	20	20	20	20	20	20	20	16.88 + 1.6 ofc.	-
525	Interest Expense	135	135	135	135	125	135	135	135	135	135	135		-
530	Advertising	55	55	55	55	55	55	55	55	55	55	55		-
521	Insurance-Keyman, etc.	299	299	299	299	299	299	299	299	299	299	299		-
524	Insurance - Business	240	240	240	240	240	240	240	240	240	240	240		-
535	Miscellaneous	40	50	60	70	80	90	100	110	120	130	140		90
536	Contributions	12	12	12	12	12	12	12	12	12	12	12		-
	TOTAL	5171	5440	5545	5706	5935	6117	6259	6470	6633	6846	7015		
	NET PROFIT (OR LOSS)	(1296)	(790)	(120)	494	1040	1633	2266	2830	3442	4004	4610		
	LABOR - MAXIMUM (No. of People)	6	7	8	9	11	12	13	14	15	16	18		
	DIRECT LABOR HOURS	1022	1227	1432	1636	1841	2045	2250	2455	2659	2864	3068		
	DIRECT LABOR COST/HR.						330							
	OVERHEAD COST/HR.						330							

Figure 3.2 Monthly profit and expense control schedule for Sample Electric Company. This general overview of how the profit and expense control is formulated serves to create understanding *only*. Tables 9.1 and 9.3 are preferred by this author for practical use in business.

5. This equals 1 percent.

6. Multiply the answer by the percentage of level of activity.

As soon as actual operating revenue and expenses are available, they should be entered. The standard figures should be calculated using the same level of operating revenues as the actual net operating revenue. Differences should be calculated between actual and standard for each expense. Where actual exceeds standard, the differences should be circled and considered negative. Where actual is less than standard, the differences should be considered positive. Each month the differences are added to, (if one is negative and the other is positive), or subtracted from, the year-to-date difference from last month's report.

Changing the Standard

When the expense has been over or under the standard for more than 6 months, or when business conditions change so much that expenses should be reduced or increased, changes can be made to reflect rising or decreasing expenses. The levels of expenses, however, should be calculated by a competent accountant and should not be changed unless you are sure that the standards cannot be met or unless conditions make them unattainable.

Changing the Activity

Should the company enter other lines of activities or increase operating revenue beyond the scope of schedule, the entire profit and expense control schedule would have to be recalculated to reflect the different circumstances. This function should only be performed by a competent accountant.

Benefits of the Job-Cost System

The job-cost system will open your eyes. It shows whether the job costs are reasonable to bring in appropriate returns on the investment and services. It shows clearly the types of jobs which are more profitable. If some type of incentive system is to be used, the savings in labor-hours can be used as the basis for the system. But before making any decision about using the job-cost system for incentive purposes, management must be very, very sure that the estimated labor-hours are reasonable.

Until positive action is taken to eliminate the negative difference and to keep operating, general, and administrative expenses at a proper level through careful control, profit and expense control will not be in full and proper effect. All employees of the company must understand that in order to maintain continous employment with the company, they must act individually and collectively to eliminate waste and unnecessary expenditures whenever possible.

This standard procedure shall come into full force and effect when signed by the general manager, and shall remain in force and effect until withdrawn by the issuing authority.

Controls for the Job-Cost System

A correct job-cost system is a must for efficient running of any business. A business may make roaring sales in a particular year and yet not make a reasonable profit—or may even suffer a loss. Thus increased sales do not necessarily mean increased profits. This happens because as a business expands, new factors in management decisions and controls come in. These new controls have to be built into the system to enable management to get the information it needs at the correct time and then to apply the necessary controls to conserve the profits.

The job-cost system builds up the cost of each and every job on the basis of labor-hours, materials cost, overhead expenses, and a reasonable profit percentage. When the jobs are completed (as well as through different stages), the actual costs are compared with the estimated costs. If the variances between the estimated and actual costs are small, they can be ignored, but if the differences are substantial, the reason must be found and corrective action taken in the future. Also, during an existing job, corrective action can be implemented through job phases using weekly or monthly job-cost reports.

Some companies initially regard the job-cost system as a waste of time and money, but sooner or later, most companies realize that they cannot exist without it.

4

Employee Handbook: An Example

This chapter gives an example of an employee handbook. It includes many elements which you may wish to incorporate into the employee handbook for your own company.

Employee's Signature

After you have read and understood this handbook, sign on the line below (and add the date), and also sign the safety pledge. Then return the signed handbook to your supervisor, and pick up an unsigned copy of the handbook for your own personal use.

_____   _____
(Signature)   *(Date)*

Employee's Safety Pledge

I pledge to do nothing unsafe, to do everything in my power to keep from being injured myself or from causing an injury to any fellow worker, and to maintain at all times an interest in the absolute prevention of accidents.

_____   _____
(Signature)   *(Date)*

Employee Handbook: Sample Electric Company

The employee handbook is intended to serve as a convenient reference to company policies and procedures for our employees. It also provides potential employees with information about what will be expected of them and what they can expect from Sample Electric Company.

When we talk about Sample Electric Company or "the company," we really mean "people" working together. Ours is a service company. We do not "make" anything. We provide our customers with a service—installation of an electrical system. So, our first concern as individuals and as a company is to meet the varied needs of our customers. Each time we fully satisfy one of our many customers, we earn for our company enhanced confidence and respect, and we also attract new customers.

For 20 years Sample Electric Company employees have shared in the responsibility of making the name of Sample stand for service in the electrical industry. As a Sample employee, you now share that responsibility.

Just as you share the responsibility for the success of Sample Electric Company, you also share the rewards which come from that success. These rewards include the satisfaction of a job well done and direct compensation in salary and benefits.

As you read this handbook, keep in mind that we earn our benefits through the excellence of our services.

History of Sample Electric Company

The company was founded by John Sample in 1962, 2 years after he was graduated from State University. He began operations by wiring houses in surrounding cities and counties. During the next 3 years the company solidified its reputation in the construction industry and expanded into commercial wiring. In 1965 the company was incorporated.

Company Profile

This section includes growth divisions and activities which may not presently be in force, but which are mentioned and explained to help employees understand the future operational effect.

Sample Electric Company is a growing company with an ever-growing need for honest, capable, ambitious people to whom we can provide opportunities for growth and advancement. Our recruitment and personnel policies have enabled us to promote all management personnel from within.

The company presently has two divisions: construction and administration. The construction division is engaged primarily in the installation and maintenance of electrical systems. The administrative division provides all necessary record-keeping, secretarial, personnel, administrative, and other support functions. Each of our offices has an organizational chart and a job-description manual clearly defining all areas of responsibility.

As growth continues, other divisions will be added in order to keep pace with the organizational needs created by expansion. A general manager will be appointed to facilitate coordination of division functions. This key management official will assume responsibility for ensuring the flexibility required to solve the dynamic problems encountered in the management of a progressive, growing company.

A marketing division will be added to provide technological assistance as well as engineering and drafting services. In addition, this division will work closely with the construction division in the preparation of bids and estimates.

An operations division is also planned. This division will be subdivided into three departments: centralized accounting, maintenance, and centralized warehouse. The centralized accounting department will maintain all company financial records, provide computer services, and prepare a variety of accounting-related reports.

The maintenance department will be responsible for motor vehicles, tools, equipment, and physical plants. It will provide transportation services, job-storage needs, and all preventive maintenance and repair work.

The central warehouse division will primarily serve the construction division by assisting with stock- and job-material purchases and by functioning as a central distributor for excess or salvaged material.

The central warehouse division manager will advise the other division managers about prices to use for the materials and tools normally warehoused in a construction division and in the control of their inventory.

The special division may operate in any geographical location. Any special project which the local division may not be staffed or equipped to handle is usually transferred to the special division, as is any large project of short duration that would interfere with the normal work load of a local division. The special division usually also performs any work that must be done outside the geographical area of a local division.

Company Policies

Employment

It is the continuing policy of Sample Electric Company to afford equal employment opportunity to qualified individuals regardless of their race, creed, color, national origin, age, or sex, and to conform to applicable laws and regulations. (See Figure 4.1.) This policy of equal opportunity covers all aspects of the employment relationship, including, for example, hiring, work assignments, training, discipline, facilities, pay rates, and other benefits.

Before you can be accepted for employment, you must read the current company employee handbook. This requirement is for your benefit as much as it is for ours. The information in the handbook tells what will be expected of you and what you can expect of us. The person interviewing you will answer any questions you may have after you read the handbook.

You may be employed as either a regular employee or a part-time employee. Before you can be considered a regular employee, you must satisfactorily complete 4 consecutive weeks of work (as a trial period, required of all employees). If you do not meet all the criteria necessary to be a regular employee, you will be a part-time employee.

AFFIRMATIVE ACTION PROGRAM
FOR
EQUAL EMPLOYMENT POLICY

To comply with the rules and regulations of Executive Order 11246 and Title VII of the Civil Rights Act, this affirmative action program has been adopted.

1. In order to conform with state and federal laws and executive orders, _____ has adopted this policy and procedure, which verifies this company's intent to implement equal employment opportunity (EEO).

   _____ does not discriminate against any person seeking employment with our company. It is our company's policy to give all applicants equal consideration for employment. All of our employees are treated equally during their employment, without regard to sex, race, color, creed, national origin, or age. Our management and supervisory personnel and also their representatives have the primary responsibility of ensuring that no person who is qualified to perform and capable of performing the required work is discriminated against in hiring, discharge, promotion, demotion, pay, compensation, and other conditions of employment because of sex, race, color, creed, national origin, or age.

2. The responsibility of the emphasis, coordination, and implementation of the company's nondiscrimination policy in employment shall be placed with the management and the company's staff. They will meet periodically to review the policy on equal opportunity.

The EEO coordinator will meet with supervisory personnel periodically and discuss the EEO policy. The attention of all personnel and each new employee is to be directed to the employment policy at the time of hiring.

At project meetings discussion of the EEO policy and programs will be implemented.

3. The company will, in all solicitations or advertisements for employees, state that all qualified applicants will receive consideration for employment without regard to sex, race, color, creed, national origin, or age.

4. The company agrees to post, in conspicuous places available to employees and applicants, employment notices to be provided by the contractor setting forth the provisions of the nondiscrimination clause.

The company will send to each labor union or each representative, or to all workers, a notice advising them of their commitments under the executive order, and shall post copies of the notice.

5. The company will furnish all required information and reports and will permit access to books, records, and accounts by the contracting agency and the Department of Labor for the purposes of investigation to determine compliance with the regulations, rules, and orders.

The EEO coordinator shall retain the records for the duration of the project or a minimum of 1 year after hire. The names and addresses of all minority applicants will be maintained so as to include information about actions taken.

6. Following the long-standing policy, continued attention shall be given to the upgrading of employees consistent with project needs and capabilities.

7. There will be no segregation of facilities or of employees at any of the project establishments, and the company will not permit its employees to perform their services at any location under its control where segregated facilities are maintained.

Figure 4.1 Affirmative action program for equal employment policy.

8. All employee benefit programs will be administered without discrimination because of sex, race, color, creed, national origin, or age.

9. All employees have a responsibility for administration of the equal opportunity employment policy and procedures.

10. Employees are encouraged to direct opinions, information, and suggestions about equal employment opportunity to the compliance coordinator. The compliance coordinator is available to discuss EEO matters.

11. The company will utilize a bona fide apprenticeship program to encourage minority employees to increase their skills and job potential through participation in such training and educational programs.

EEO Coordinator

Figure 4.1 *(Continued)*

Attendance

The company places high value on good attendance. When you must be absent, give your supervisor as much notice as possible. If the absence is unexpected, notify your division office within a half hour of your normal time to begin work.

Reporting Changes

Social Security Records.   If your name changes, through marriage or for any other reason, you must fill out a form to update the social security records. Ask your field supervisor to get the appropriate form for you.

Federal Tax Exemptions.   When a change occurs in the number of your dependents, you must fill out the required federal government withholding form to have this change reflected in your paycheck. Reporting an increase in the number of dependents is optional; reporting a decrease is required by law. Notify your field supervisor if you wish to make a change.

Status.   An employee information form must be filled out if you change your address, home telephone number, or marital status. If there is a change in your military status, notify your division office.

Company Property

Tools and Equipment.   Company-owned tools and equipment furnished to you are for your use in the performance of your job. No other use of company tools is allowed. You are responsible for the proper security, use, and on-the-job maintainance of company-owned tools and equipment assigned to you.

Company Vehicles.   The following rules apply to all employees.

1. Employees are to use company-owned vehicles only while working and during working hours, unless otherwise authorized.
2. No unauthorized use of company vehicles will be permitted.
3. Employees who are authorized to take company vehicles home shall drive the company vehicles only directly to and from work during nonworking hours.
4. Company-owned vehicles may not be used for transporting persons other than employees of this company at any time.
5. Personal vehicles are not to be used on company business unless authorized.
6. All use of company vehicles must be reported on the individual vehicle record and must include daily starting and ending odometer readings, destinations, driver's initials, and record of refueling.
7. Violations of these policies will cause the employee to be subject to termination and discipline procedures.
8. Sample Electric Company will not pay for employees' traffic violations.
9. Sample Electric Company will not be legally liable for any traffic violations by its employees.
10. The accounting department does monitor vehicle expense through mileage and gas checks, etc.

Unused Material.   All unused material, including material that becomes scrap as a result of the work being performed, is the property of the company and must be returned to the division warehouse. All company employees are bonded by a company bond, which covers material, scrap, tools, supplies, and money. Any incidents involving improper use of unused material by employees will be reported to the bonding company for investigation and proper action.

Authorized Leave of Absence

If you must be absent for 1 work week or more, you should request an authorized leave of absence from your division manager. Each request will be granted or denied after a review of the individual case. It is company policy not to grant an authorized leave of absence to an employee for the purpose of accepting temporary employment elsewhere.

Pregnancy

Employment decisions involving a pregnant employee will be based on the employee's ability to perform the customary duties of the position safely and efficiently.

Military Leave

It is company policy to comply with both the letter and the spirit of all laws pertaining to military leave. Any regular employee who enlists or is drafted into military service will be granted a military leave of absence. If within 90 days after completing military service under honorable conditions, the employee returns to regular employment with Sample Electric Company, the military time may be considered continuous satisfactory employment.

Part-Time Employment

Part-time employees will not be eligible to earn paid vacation or paid holidays, nor to participate in the group insurance plan (described in another handbook). Part-time employees who become regular employees will be given credit for all satisfactory service in the calculation of eligibility for the above benefits (not including insurance calculations).

Related Work

Employees are not permitted to do any work for other employers that falls within the work scope of Sample Electric Company (installation or maintenance of an electrical system).

Unionization

Sample Electric Company strongly endorses the philosophy that individual consideration in employee-supervisor relationships provides the best climate for maximum development of the individual and for attainment of the goals of both the individual employee and the company.[1]

We do not believe that union representation of our employees would be in the best interests of either the employees or the company.

We acknowledge the right of our employees to join a union if they wish. When they do so, however, they assign to the union the right to represent them as sole agent in respect to most matters relating to their conditions of employment. For this reason, we believe that the mere presence of a third party between company and employees takes away some of the rights of the individual and prevents individual employees from realizing their full potential. We sincerely believe that any third party could seriously impair the relationship between the company and employees, and could retard the growth of the company and the progress of the employees.

We are certain that our personnel policies and programs for pay, benefits, and services provide the employees with the best possible opportunities for individual growth and attainment of personal goals.

Safety Policy

It is the policy of Sample Electric Company to fully comply with the latest Occupational Safety and Health Act (OSHA) and Construction Safety

---

[1] The above is for open shop and is naturally just the reverse for union contractors.

Act. The intent of this policy is to provide our employees with the safest working conditions possible.

We realize that our employees are our most important asset and that their safety is our greatest responsibility. Therefore, we not only ask but require your full and complete cooperation in complying with the provisions of OSHA. This means specifically that:

1. You are responsible for supplying your own personal safety equipment, which must comply with the latest Occupational Safety and Health Act and Construction Safety Act. ("Personal safety equipment" consists of approved head protection, eye protection, and all articles of clothing, including footwear.)
2. You can be dismissed for a safety violation.

It is not our object to increase the difficulty of employees' jobs nor to penalize them in any way; rather, all employees of the company must work together to prevent accidents. Remember that construction injuries are always costly to individual workers and often disastrous to their futures and to the security of their families.

The company also requires full compliance with OSHA and the Construction Safety Act by all subcontractors and suppliers.

Safety Rules

The safety rules below are of the utmost importance and must be obeyed by all employees at all times.

1. Always keep in mind the safety of your fellow employees, the public, and yourself.
2. Work within your physical limitations and the limitations of your tools and equipment.
3. If you are physically or mentally unable to perform your duties, inform your supervisor immediately.
4. Before beginning any job, determine the safest way to accomplish the task, and which tools and equipment you should use. If any piece of equipment or any procedure looks unsafe, ask your supervisor for assistance.
5. High-quality hard hats (Jackson SC3 or equal) shall be worn under the following conditions:
   a. When there is danger of falling objects
   b. In roped-off areas
   c. While working on high voltage
   d. While installing a bus switch on an energized bus
   e. When on a moving elevated platform
6. Always keep the work and stock area clean; leave all job sites clean and in a presentable condition after the job shift.
7. Keep floor or vertical lifts, person lifts, and scaffolds clean and clear of nonessential materials and tools.
8. If working on an electrical circuit or on equipment that could accidentally become energized, lock out and tag the main switch.
9. Use extreme caution when working on energized equipment. Lock or

lockout tags can only be removed by the persons who placed them, and then only if no one is in a position to be injured.

10. Before using a ladder, be sure the following conditions are met:
    a. It is in good repair.
    b. It is long enough for the job.
    c. It is equipped with either safety shoes, spikes, or spurs, depending on the job.

11. While using a ladder, keep the following precautions in mind:
    a. When a ladder is placed near a door in such a position that opening the door could displace the ladder, the door must be locked or guarded.
    b. When additional safety equipment is required, the ladder must be tied for support, or held by another employee.
    c. Never work from the top step of a ladder.
    d. Do not use metal ladders when working on energized circuits or equipment.

12. When on an elevated surface, be sure proper equipment is used in a safe manner. Outriggers must be set before using a person lift or scaffold, if equipped for outriggers.

13. No passengers will be allowed on a vertical lift, or other scaffolding; use extreme caution in low-clearance areas.

14. Rope off the work area when working overhead (OH), except in hard-hat areas.

15. After installing outlets, check them with a tester for correct wiring before turning the site over to the customer.

16. If in doubt about the advisability of making a connection or performing a procedure, stop and ask your supervisor for assistance before proceeding.

17. Do not stand or walk on bus or light fixtures.

18. Identify (mark clearly) all disconnects, bus switches, and circuits after installation.

19. Never attempt to operate any equipment unless you are both qualified and authorized.

20. If you enter a contaminated atmosphere, be sure to take the necessary precautionary measures.

21. Always give fellow employees the benefit of your experience. Do not become complacent or set an example that may encourage less experienced employees to take hazardous shortcuts.

22. Portable electric tools must be of the grounded type or must have the approved type of double insulation.

23. Use special self-protective equipment whenever you enter trenches, plating areas, pits, manholes, or vaults, and whenever you climb towers. If you have questions about the need for or availability of the appropriate equipment, ask your supervisor.

24. Inspect rubber goods or hot sticks before you use them.

25. In case of sickness or injury, report at once for first aid and notify the company as soon as possible. Fill out an accident report and send it to the office within 24 hours.

26. While driving a vehicle, use seat belts. Obey all traffic laws, as well as the unwritten rules of courteous driving.
27. Do not wear contact lenses while working. Exposure to dust, chemicals, or a flash or arc is dangerous when wearing contact lenses.
28. Put appropriate warning signs and barriers in place promptly when covers of manholes, handholes, or vaults (vented or unvented) are removed. No entry shall be permitted until forced ventilation has been applied for at least 5 minutes. An adequate and continuous supply of fresh air must be available while working in manholes. Whenever an employee is working in a manhole, a second employee shall be available in the immediate vicinity to render emergency assistance if required.
29. While working on the roof of a building, 10 feet or less from the edge of the roof, wear a safety harness or safety belt with a lanyard attached to an appropriate anchorage.

Violators of these rules will be subject to the discipline and termination procedures.

Accident-Control Program

All employees are responsible for contributing to the success of the accident-control program. All employees have a responsibility to themselves for their own safety, but they also have a responsibility to their family, fellow workers, the community, the customers, and their employers.

In the performance of your duties you will be expected to observe safety rules, as well as instructions relating to the efficient performance of your work. A safe and efficient construction operation is achieved only when all employees are safety-conscious and keenly alert mentally and physically.

It is your responsibility to:

1. Notify your supervisor immediately of any conditions or practices that may cause personal injury or property damage.
2. Report all accidents and injuries to your supervisor immediately.
3. Submit suggestions for safety and efficiency.
4. Know your exact duties in case of fire or other catastrophes.
5. Observe all safety rules and regulations and make maximum use of all protective equipment.

Intoxicants and Drugs

Employees are not permitted to have in their possession, to use, or to be under the influence of alcohol or drugs (except as prescribed by a doctor for temporary illness not affecting their job performance) at any time when they are on company property or job sites, or in a company-owned vehicle. This policy is absolutely necessary for your protection as well as for the protection of your fellow workers, other tradespeople, and the customer.

Solicitation

Solicitation and distribution of literature by employees on company property or on company job sites during working time are prohibited if these activities interfere with work in any way. Solicitation and distribution of literature by nonemployees on company property or on company job sites are prohibited at all times.

Complaints

The management of Sample Electric Company recognizes that the accumulation of unspoken, unanswered complaints results in dissatisfaction and destroys what otherwise could be a pleasant work relationship.

Therefore, when a problem does occur, it is to everyone's benefit to assure a prompt and fair solution. That is the purpose of the following "open-door" procedure:

- When you have a problem or complaint, you should discuss it with your supervisor at the earliest convenient time. It is part of your supervisor's responsibility to try to correct situations which cause problems and complaints. Therefore, it is very possible that your complaint can be settled by the two of you to your mutual satisfaction.
- If you and your supervisor do not reach a solution, you may then ask for an appointment with your division manager. Your supervisor should arrange this appointment without prejudice to you; making such arrangements is also part of his or her responsibility.
- If you do not feel that the problem has been properly resolved by your immediate supervisor or the division manager, or if the problem is such that you wish to have a private consultation with someone other than your supervisor or division manager, you should ask for an appointment with the company personnel director. You should, in fact, be aware that you may call on and consult with the personnel director concerning any problems that you feel warrant that person's attention.

Suggestions

The company is always receptive to suggestions for better ways to serve our customers. As an interested, knowledgeable employee, you are likely to develop some new ideas. Besides giving you the personal satisfaction that comes from creativity, the ideas you submit may possibly bring you cash rewards. Suggestions should be submitted to the executive committee. Suggestions may be submitted either under your own name or anonymously. Please identify your suggestion with a code number if you wish to remain anonymous.

Grievance Procedure

The grievance procedure for Sample Electric Company is shown in Figure 4.2.

Resignation

A minimum of 2 weeks' notice (10 working days) must be given, in order to leave in good standing.

Discipline and Termination

Our company will maintain a personal file on each employee. Employees can accumulate three written warnings in their file. (See Figure 4.3.) This policy constitutes the first written warning; the fourth written warning will be cause for termination. The classes of violations are listed below. Warnings are issued for class 1 violations. Class 1 violations may also be cause for termination, at management's discretion. Class 2 violations will be cause for immediate termination. Class 1 violations are as follows:

1. Poor productivity
2. Poor workmanship
3. Tardiness
4. Quitting early
5. Performing unauthorized work
6. Being absent without notifying the employer
7. Abuse of coffee privileges
8. Intentional slowdown of work

GRIEVANCE PROCEDURE

1. Take up any grievance with your immediate supervisor. He or she has 3 days to give you a satisfactory answer.

2. If you feel that the 3-day period has passed, that no satisfactory answer is given, or both, you may request a meeting with your supervisor and his or her superior. A 3-day waiting period is again required, but with this meeting a solution to the problem must be reached.

This procedure is standard and becomes law when signed and dated by the president.

_____   _____
Date   President

Figure 4.2 Grievance procedure.

```
┌─────────────────────────────────────────────────────────────────┐
│                  EMPLOYEE WARNING NOTICE                          │
│                                                                   │
│                                               1st Notice          │
│                                               2d Notice           │
│                                                                   │
│  Name_____    Date_____│
│                                                                   │
│  Dept._____    Clock No._____ │
│  ---------------------------------------------------------------- │
│                                                                   │
│  Nature of Violation                    REMARKS                   │
│   ☐   Carelessness        _____ │
│   ☐   Absence             _____ │
│   ☐   Disobedience        _____ │
│   ☐   Lateness            _____ │
│   ☐   Safety violation    _____ │
│   ☐   Poor housekeeping   _____ │
│   ☐   Poor conduct        _____ │
│   ☐   Defective work      _____ │
│   ☐   Poor attitude       _____ │
│                                                                   │
│  _____      _____     │
│  Signature of Foreman or Supervisor   Official Signature          │
└─────────────────────────────────────────────────────────────────┘
```

Figure 4.3   Employee warning notice.

9. Violation of company policies in regard to EEO, safety, tools, work rules, vehicles
10. Violation of customer rules (no smoking, etc.)
11. Neglect or abuse of company tools, equipment, or vehicles
12. Concealment of errors and mistakes
13. Unauthorized gathering or assembly

Class 2 violations are as follows:

1. Theft of company or customers' property
2. Insubordination; refusal to perform assigned work
3. Fighting on the job
4. Sleeping on the job
5. Possession, use, or being under the influence of alcohol, marijuana, other illegal drugs, or unauthorized drugs
6. Possession on the job of weapons, explosives, cameras, recording equipment, or transmitting devices without consent from the company
7. Dangerous work practices, horseplay, etc.

8. Destruction or defacement of company or customers' property
9. Falsification of records, time cards, etc.
10. Moonlighting of electrical work
11. Representing the company without authorization
12. Harassment of other employees

Administration of Rules and Policies

As an electrical contracting firm, Sample Electric Company recognizes that our greatest asset is our employees. As a business, it is our aim to get the most from all our assets, especially our people. In order for employees to perform at the highest level, rules and policies must be clear. It is up to management to set the policies and make the rules. Our rules are fair and will be used consistently and without bias to improve morale and productivity. Our rules and policies make the limits clear to employees. Policies on tools, vehicles, and materials increase profits through more efficient use of assets and fewer losses from damage and theft. The policies and rules set forth in this manual are just as much for your benefit as for the company's.

This set of policies and rules is nothing new; we have had them all along. We foresee no change in these policies and rules unless the rules are broken. Whether you are a new employee or have been with the company for some time, you are required to read and ask questions about the company rules and policies until you understand them, and then sign in the proper place at the beginning of this manual to indicate that you have read and understood the rules. Arrange for a short meeting with your supervisor for the purpose of asking questions about any rules that you do not understand upon first reading.

Company rules and policies can be an effective management tool—or they can cause many problems. Problems occur when rules are not used, or when they are used inconsistently or unfairly. It is our intent to prevent this by the following procedures:

1. Only supervisory personnel are allowed to write warnings or terminate employees, and then only after consultation with the manager.
2. No warnings are to be given or discharges made without an investigation of the facts by all involved.
3. No employee shall be disciplined for breaking rules which other employees have been allowed to break before.
4. When new rules are to be enforced, all employees must receive written notice.
5. Discipline is matter of corrective procedure, not punishment.

On the Job

Before You Start

If you have had previous electrical construction experience, you will be given a briefing on the use of hand tools, safety, and general company production procedures before you begin work. This briefing will be given by either the person who hires you or the person with whom you are assigned to work.

Tool Requirements

All mechanics, apprentices, and helpers shall furnish and have the following tools in their possession at all times during working hours.

1 tool pouch and belt

1 pair needle-nose pliers

1 pair side-cutter pliers

2 pairs channel lock pliers

1 large screwdriver

1 device screwdriver

1 small screwdriver

1 Phillips screwdriver

1 folding rule

1 pocket level

1 pocketknife

1 pencil and notebook

In addition, all production employees must furnish and have in their possession a safety hat and a pair of safety glasses or goggles at all times during working hours.

The following is a representative but not complete list of additional tools that must be furnished by employees if required by their immediate supervisors.

1 toolbox

1 Stakon tool

1 Yankee screwdriver

1 set nut drivers

1 Rawl drill for No. 8 and No. 12 bits

1 Rawl drill or Star drill, $\frac{1}{2}$-, $\frac{5}{8}$-, or $\frac{3}{4}$-inch

1 hammer

1 rattail file

1 flat file

1 hacksaw

1 small cold chisel

1 large cold chisel

1 center punch

1 set drill bits, $\frac{1}{16}$- to $\frac{1}{2}$-inch, by 32ds

1 pair KO cutters, $\frac{1}{2}$- to $1\frac{1}{4}$-inch

1 adjustable wrench, 12-inch

1 keyhole saw

1 racket brace

1 wood bit, 1-inch

1 pair tin snips

1 set Allen wrenches, to ½-inch

1 set tap wrenches, ⁶/₃₂-, ⁸/₃₂-, ¹⁰/₂₄-, ½-, and ¼-inch

1 tape line, 50- or 100-foot

1 chalk line

1 plumb bob

1 flashlight

1 pipe wrench, 14-inch

1 fish tape, 100-foot

1 blowtorch

1 ladle

2 hickeys, ¾-inch

1 hickey, ½-inch

2 benders (electrical metallic tubing, or EMT), ½- and ¾-inch

1 thin-wall crimping tool: ½-, ¾-, or 1-inch

1 level, 24-inch or longer

1 socket set

1 set box and open-end wrenches

Tools owned and furnished by you may be replaced by the company if you wear them out while working for the company. Your field supervisor must approve each replacement.

Transportation to Work

It is your responsibility to provide your own transportation to work. In most cases your job foreman is assigned a company truck to carry tools and material to the job site when necessary. You may make arrangements with your job foreman to ride to work if it is convenient.

If you are asked to drive your personal vehicle for company use, you will be paid the current company mileage rate per mile.

Where to Report

Each employee will be assigned a job site or division office to report to and a time to begin work. If you are required to change job sites during the normal working hours, your travel time will be charged to each job proportionately or as directed by your job foreman.

Work Hours and Rules

Electrical contracting is very competitive. This company needs 8 hours of work for 8 hours of pay from all its employees in order to be competitive and provide employment for its people. A spirit of teamwork and cooperation increases productivity. Constructive suggestions are always welcome.

Your normal working hours will be approximately 8 hours per day, Monday through Friday, but your supervisor may require you to work overtime when it is deemed necessary. Any hours over 40 per week on your time sheet will be overtime, and you will be paid 1½ times the applicable rate. There will be no overtime payments on fringe benefits, nor on government jobs or any jobs operating under U.S. Department of Labor guidelines.

Each day, ½ hour without pay will be allotted for lunch. Your job foreman will tell you when your lunch break begins and ends.

Two 10-minute rest breaks with pay, one in the morning and one in the afternoon, are allotted. Your job foreman will determine when these breaks are to be taken.

You must follow the work rules listed below:

1. Starting time is 8 a.m.; quitting time is 4:30 p.m. Exceptions may be made if caused or directed by special situations, such as starting and quitting times on U.S. government bases.

2. Be on the job site and ready to work at starting time.

3. No employee shall leave the job site before quitting time without advising our office and obtaining permission.[2]

4. Pick-up and clean-up time shall be held to a minimum.

5. Employees must remain in their assigned work area unless otherwise instructed by their supervisor.

6. There will be no organized or scheduled coffee breaks. Employees may have a 10-minute coffee break at their work station only when specifically allowed by the employer. Coffee breaks will not be allowed to interfere with job progress. Abuse of this rule will result in the loss of all coffee-break privileges.

7. Lunch breaks will be ½ hour, from the time work stops until the time work resumes.

8. Personal telephone calls are not allowed except in emergencies.

9. No radios or TV sets are allowed on the job.

10. Before moving from one job site to another, employees must notify their immediate supervisor.

11. Violations of these rules will cause employees to be subject to discipline and termination procedures.

Time Sheets

Time sheets should ideally be turned in as soon as you have finished work for the week. Time sheets *must* be turned in to the division office by

---

[2]Or unless field supervisor gives permission.

8 o'clock Monday morning so that your paycheck will be ready for you by quitting time Friday.

You must fill out and sign your own time sheet each week. If you need help your immediate supervisor will help you.

Our time sheets are not simple, but if you begin a new one each Friday and enter the job number or work-order number, the subcode, and the hours worked each day, you should not have any problems. If you and your supervisor cannot agree on the information entered on your time sheet, do not sign it. If you believe your time sheet to be incorrect, ask your supervisor to let you discuss the problem with your division manager.

Pay Periods

The company's work week begins on Friday and ends on Thursday. Paychecks for the work week ending Thursday will be distributed no later than quitting time the following Friday.

What You Earn: Wages, Job Classifications, and Benefits

While working for Sample Electric Company, you will earn compensation in a number of forms. One very important form of compensation is wages, which are closely tied to job classifications. But your pay is only part of what you earn. Also included in your compensation are a wide range of valuable benefits and services which save and stretch your earnings. Sample Electric Company's benefits program is a security package you earn through your service. It is designed to put your mind at rest about everyday worries and also to help you get the most from your leisure time. We will discuss the two forms of compensation—wages and benefits—in the next few pages.

Two other possible additions to your compensation that won't appear in your paycheck are our pension plan and our group insurance plan. These two forms of compensation are explained in other handbooks. For more information ask your field supervisor.

Your Wage Rate

Job Classifications. Jobs are classified according to the technical knowledge and experience necessary to perform each job. Each classification has a wage-rate limit: the highest hourly rate that can be paid to persons in that classification. The production job classifications are laborer, helper, apprentice, mechanic, service worker, work-order foreman, job foreman, work-order superintendent, job superintendent, and project superintendent. (See Figure 4.4 for examples of job designations and qualifications.) Not all these classifications are presently used.

How Your Rate Is Set. Your wage rate lies somewhere within the range for your classification. Ordinarily, you start at an average rate for your

---

JOB DESIGNATIONS AND QUALIFICATIONS

A. TRAINEE HELPER: Should preferably be a high school graduate. Must have a driver's license. Must be able to read and understand road maps thoroughly.

B. HELPER: Must know all the tools and equipment. Must be able to differentiate between the different electrical components and materials. Responsible for bringing back all the materials and tools when the work is over. Should also be able to fill in the time and material card. Responsible for stocking all the things required for working at the site and keeping the truck in proper order. Also must run house circuits, and know symbols for house wiring.

C. ELECTRICIAN I: Must be familiar with the plans and specifications. Must know wiring methods and materials thoroughly, as shown in Chapter 3 of the current *National Electrical Code®* (NEC).* Must know the different sizes of wires, and when and where to use them.

D. ELECTRICIAN II: Must be familiar with the NEC tables which relate the number of conductors in certain size conduits. Must be able to lay out regular wiring and electric heating circuits; install services and vacuum systems; wire heating and cooling units; etc.

E. ELECTRICIAN III: Must be thoroughly up to date on all aspects of the electrical codes, and must be able to solve most problems independently, without any supervision.

F. ELECTRICIAN IV: Must have electrician's cards from the county and city. Must be up to date on single-phase, star, and delta three-phase systems. Must know how to balance the phases for equal load distribution. Must be familiar with motor circuits and controls for 277- to 480-volt systems. Must be able to work out a commercial wiring plan and control the job from start to finish.

Pay scales are subject to change depending upon the market conditions and business profits.

If several workers are on a truck together, and none of them is defined as a helper, the tool and stock responsibility shall be shared.

Advancements and promotions are to be made in accord with all the above, plus a standard written or oral test designed by your supervisor. You should set a deadline for your cards and work toward getting them.

*National Fire Protection Association, *National Electrical Code*, Chapter 3, Wiring Methods and Materials, NFPA 70, NFPA, Quincy, Mass., 1986. (Triennial publication.)

Figure 4.4   Job designations and qualifications.

classification, unless you have some unusual qualifications. From then on, how well you do your job determines your rate, within the range for your classification. Your job foreman, field supervisor, and division manager set your rate and make any changes. Their decision is guided by our wage policy.

Rate and Classification Review.   Adjustments in wage rate, classification, or both can be made at any time during the year, when justified. However, to make sure that every employee is considered, formal rate and classification reviews take place periodically during the year.

Pace of Increases in Wages.   How well you do your job determines how far and how fast you move up in your wage range. *There is no automatic rate of increase.*

Promotions. When you earn a promotion or a higher classification, you may or may not receive a rate increase. However, the range of your wage rate will become wider.

Benefits and Services

Most of the benefits and services presently in effect at Sample Electric Company are discussed below. Pension plans and other benefits mentioned in this manual may or may not presently be in force. Check with your supervisor.

Vacations. All regular employees are eligible to begin earning paid vacation when they begin work. Paid vacation will be earned at the rate of 3⅓ hours per month of satisfactory employment.[3]

You earn paid vacation each year from the time you become eligible until the end of the corporate year. The corporate year runs from October 1 to September 30.

You will be compensated for earned paid vacation during the corporate year immediately following the year it is earned. Compensation for earned paid vacation will be at your regular base rate. If you do not take your earned paid vacation, you will be paid for it (the last month of the corporate year). This payment must be requested by your division manager or supervisor.

If you are eligible for paid vacation and quit working for the company without notifying your supervisor, or if you are fired for any reason that makes you ineligible for unemployment compensation, you will not receive any unused or unpaid vacation pay.

If you are eligible to be paid for a holiday that falls during the period that you are on vacation, then that day will count as a paid holiday and not as a day of paid vacation.

If you are eligible for a paid vacation and are absent because of circumstances beyond your control (for example, illness), then you may request that all or part of the time that you were absent count as paid vacation time. Under no circumstances will paid vacation time be taken before it is earned.

Holidays. Sample Electric Company will observe the following holidays:
Independence Day.
Labor Day.
Thanksgiving Day.
Christmas Day.
New Year's Day.
After 5 years of employment, add birthday.
After 8 years, add December 26.
After 10 years, add January 2.

---

[3]Not to exceed 3 weeks.

Any employee properly classified by the division manager as a service worker, work-order foreman, job foreman, work-order superintendent, job superintendent, or project superintendent, and all regular employees with 1 or more years of satisfactory cumulative service, will be paid for the holidays listed above. The compensation will be for 8 hours at the employee's base rate of pay.

As a production employee, you may, depending on the job requirements, be required to work on holidays. If you work on a holiday that you are eligible to be paid for, you will be paid the applicable rate for the time you work, plus 8 hours of holiday pay at your base rate of pay.

If any of the above holidays fall on Saturday, they will be observed on Friday. If any of the above holidays fall on Sunday, they will be observed on Monday.

Social Security (FICA).   Retirement income from this government plan will begin when you reach the retirement age specified in the law. If you become disabled, you may qualify for Social Security disability benefits. See your local Social Security Administration office for details.

Workers' Compensation.   If you are injured on the job, you will receive medical expenses and loss-of-income compensation through this plan. The company pays the premiums for this plan.

Unemployment Insurance.   The company pays for this plan, which provides you with cash to help meet expenses if you become unemployed because of a lack of work.

Veterans Administration Benefits.   Employees who have completed an active military tour of at least 181 days of active duty since January 31, 1955, and have registered with the Veterans Administration (VA) Apprenticeship Council as an electrical apprentice working for Sample Electric Company, may be eligible for VA benefits. If you think you may be eligible, ask your division manager or superintendent.

Dues.   The company will pay the annual dues for a state electrical contractor's license if you have one while employed by the company.

Training.   You will be paid for any meetings or training sessions you are required to attend. If you voluntarily attend a meeting or training session, you will not be paid for attending it.

On-the-job training is expected of all employees. In addition, you are encouraged to attend schools, workshops, or training sessions that will contribute to your advancement in the company. Division managers may authorize reimbursement of tuition and required materials upon satisfactory completion of such training; the division manager's tentative authorization for reimbursement must be secured before the course is begun.

Career Opportunities

As a Sample Electric Company employee, you share in the positive aspects of our continuing growth and progress. As an employee of a promising company in a leading industry, your career opportunities are exceptional. Whenever the company increases its share of the present market, tries new techniques, offers new services, or expands geographically, your job opportunities multiply.

Promotion from Within

Sample Electric Company policy has always been that promotions go to qualified employees whenever possible. Only when the necessary skill and training cannot be found among present employees are new employees hired. This policy ensures that, if you have ability and ambition, you will be considered for positions of increased responsibility.

Training for the Future

The surest path to advancement is to become thoroughly competent in your present job. Competence comes primarily from a continuing day-to-day effort to learn your total job and to become more effective in doing it. You are also encouraged to take advantage of any training or educational opportunities available to you that may speed your advancement.

Apprenticeship Program

The Sample Electric Company has established an electricians' apprenticeship program that is approved by the state department of labor. The apprenticeship course takes 4 years to complete and consists of on-the-job training and a correspondence course. Graduates of the program are highly regarded and in great demand in industry. In addition, the company pays each graduate a cash bonus.

If you have little or no formal training in electricity and construction, and are interested in a career opportunity, you should see your field supervisor about the electricians' apprenticeship program.

Estimating and Bidding

In order to make money in the electrical contracting business, (1) you must get the volume you need within a given time period, such as quarterly or annually; (2) that volume must be achieved within the estimated cost; and (3) you must collect all the money you can as swiftly as you can. The volume you need is in direct relationship to your cost or breakeven point plus a profit.

The Role of Estimating in Financial Planning

The inflow and outflow of money are of high concern to financial management. The trick is to trap inflow money and keep as much of it as possible from becoming outflow money. When inflow is greater than outflow, all is rosy. Achieving this feat is difficult in most cases, and impossible in some. After a very hard year of work, an owner may receive no return on investment.

The question is, where do estimators fit into all this? Is there anything they can do to improve the situation? The answer, in no uncertain terms, is that there had better be something they can do! As a matter of fact, they are part of the outflow—part of the overhead. But what exactly, is the role of the estimator?

The Role of the Estimator

One of the estimator's responsibilities is to utilize the bid document to produce prices low enough to get the job, but high enough to make a profit. Here is where the inflow of money for the electrical contractor has its inception.

The estimator's role is to be the "mayor" of money inflow. It is only when we view the estimating department in this light that we get a picture of the vital role the estimator plays.

In small companies, in which the estimator may decide which jobs to bid on, he or she is in fact responsible for the financial well-being of the company.

The Science of Bidding

It is hazardous to your company's health to become "bid-happy"—that is, to grab for every job you see and to bid just to be bidding. A safer practice is to approach bidding as a science. This means studying the bidders' lists; knowing the work load of your competitors and what they are likely to do; and evaluating your chances before you bid to be certain you are not wasting your time.

The Relationship between Estimating and Bidding

Estimating is a crucial part of bidding strategy. It is when you are formulating the estimate that you must be sure that your cost is adequate and that all avenues of recovery have been included. You must set estimating goals; you must set parameters for job sizes, geographic areas in which you'll work, and types of jobs; and you must have sufficient backup personnel to handle the jobs on which you bid.[1]

Negotiated Contracts

Negotiated work is naturally more profitable. In order to be in a position to negotiate contracts, you must have spent a long time developing the ability to deal with others on the basis of rapport, and you must have an excellent work record and a name for quality work and timely completions, with reasonable bonding power. Also needed is the ability to finance your work. Practically every successful electrical contractor has two or more general contractors (GCs) or owners who simply call and ask for a budget price, and then will negotiate doing the work.

Suppliers

It is very important to establish correct posture with at least three suppliers and, if at all possible, with several manufacturer's representatives. Supply houses generally have at least three representative price levels. You may not be getting the best possible price, for any one of several reasons. One way to double-check is to communicate with a known competitor after a bid neither of you obtained. Through

---

[1] For outside help in electrical management consulting and financial management, drop a note to C. L. Ray, Jr., 6949 Connie Drive, N.W., Roanoke, VA 24019, and ask for information.

trade organizational meetings, etc., I have always been able to develop enough rapport with competitors to obtain this information.

What suppliers look for from you:

1. Prompt payment

2. Fulfillment of what you agree to do

3. Proper purchase order with instructions

4. Curtailment or prevention of returned merchandise

What you want and must have from suppliers:

1. The lowest quotes

2. Timely quotes

3. Timely and correct shop drawings

4. Timely job-site deliveries

5. Discounts

Sometimes all the profit a contractor may make is accounted for by discounts received from suppliers. Most supply houses will give 2 percent discounts per month for prompt payment of accounts, which adds up to 24 percent annually. Even if you borrowed money at 12 percent to make prompt payments, you would still make a 12 percent profit. Taking advantage of discounts is one sure way to receive the lowest quotes from suppliers. Attend supplier functions, to get to know a number of suppliers. Call your suppliers' sales representatives, have lunch with them, and try to develop friendly and cooperative working relationships.

In selecting suppliers, keep in mind that the highest-priced supplier may actually be the least expensive one. Here is an example: Supplier X quotes you $20,000 on the material you need, can deliver all shop drawings and guarantee approval in 5 to 10 days, and can deliver all materials in 30 days. Supplier X thus is obviously cheaper than supplier Y, who quotes $15,800, cannot deliver shop drawings for 15 to 20 days, cannot guarantee approval, and cannot deliver until 60 days after placement of your order. *Time is money.*

Once the shop drawings have been approved, the field superintendent and foreman must have a set marked "approved." From this point on, you're in control of what you accept at the job site. If the materials delivered are in direct conflict with the materials approved on the shop drawings, *do not accept them.* The enormous error of allowing delivery of incorrect materials will create the following nightmare: Receipt of the wrong materials immediately starts aging of your accounts payable (AP) with the supplier. Say, for instance, that materials were delivered on the 1st of the month. The general contractor's superintendent and inspector check out the materials on the 15th and discover they're the wrong materials. In the meantime, your cashflow is contingent upon your billing for these items. By the 20th it is quite conceivable that some of these items may have been installed. Once the error is discovered, however, you ask the supplier to pick up the material. More than likely, the supplier will not correct or stop the aging of your account. You are faced also with the necessity for pulling out the installed materials. Double labor! There is no way this job can be profitable.

As this example shows, the proper handling of shop drawings will make or break the contractor's job budget and cashflow. Two or three jobs of this nature will put you out of business. It is also important *never to accept quotes from suppliers unless they give you an itemized breakdown*, particularly if you already have the job. Usually, you can accomplish this by letting the salesperson know you do intend to order, then making the itemization a precondition for issuing the purchase order (PO). The reason for this is to prevent the supplier from front-end loading your billing when materials begin to arrive at the job site. For example: A supplier quotes you 50 fixtures at a lump-sum price of $400 each. Later on, you find out he shipped you 10 fixtures and billed you for $7500, or $750 each. He should have billed you for 10 × $400 or $4000. Why allow the supplier to use $3500 of your money? You would not know what had happened without the itemized breakdown.

The Prebid Conference and Site Visit

The prebid conference usually takes place during the site visit. Some contract documents make the site visit a requirement for bidding. It is during the site visit and prebid conference that you meet your competitors. This is your opportunity to find out who is bidding, to ask questions, and to establish rapport with general contractors.

Dealing with General Contractors

In dealing with general contractors, never ask them what their quote is from other subcontractors, unless you're sitting in their office and can see on paper what they are telling you. My reason for advising against this practice is that hearing such a quote could influence your thinking. Some subcontractors belong to a bid depository for the purpose of obtaining bona fide information on their competition after the bid opening. General contractors have a history of using subcontractors to make profit and finance their job. GCs are entitled to a profit, but not at the expense of their subcontractors.

Insurance Requirements

Insurance requirements are usually specified in the contract documents. Before starting to work, be certain that all the necessary insurance requirements have been met. The assistance of someone experienced with insurance can prove very valuable.[2]

Internal Bookkeeping and Cost Categorizing

In my work with contractors on their internal bookkeeping, I have found that very few have their costs properly categorized. This distorts their ratios relative to the electrical industry.

True, the bottom line is "How much did I recover?" But, to know your financial ratio and why the banker did or did not make you a loan, cost figures must be in the proper category to make industry comparisons.

What to Do After You Get the Job

If you are the low bidder and you get the job, what do you do next? (1) Check the work and paying record of the contractor or owner; if they are good, (2) sign the contract. (3) Secure a performance and payment bond. (4) Use the proper purchase-order form to buy the materials. (See Figure 5.1.)

---

[2]For professional bonding and insurance service, write to Tom Brown and Company, Inc., P.O. Box 19293, Washington, D.C. Tell him I recommended him; you may get a special rate.

NOTIFICATION TO SUBCONTRACTORS, VENDORS AND SUPPLIERS

The _____ is a federal government contractor operating under a policy of equal opportunity and an affirmative action program in compliance with Executive Order 11246, as amended, and its implementing guidelines. Since our government-related purchasing from your firm exceeds $10,000 annually, we are required to ask you to reply to the following questions:

1. Do you have an equal opportunity policy statement which prohibits discrimination by race, sex, color, creed, national origin, and age?

   Yes _____ No _____

2. Have you developed a certification of nonsegregated facilities?

   Yes _____ No _____

3. Have you filled standard form 100 (EEO-1)? (Not required if there are fewer than 100 employees in your work force.)

   Yes _____ No _____

4. If you have 50 or more employees and the dollar volume of our government-related purchasing exceeds $50,000 annually, you are required to have a written affirmative action program on file. If your facility is part of a corporation operating two or more facilites, any one of which provided goods or services valued at $50,000 annually to the federal government, and your corporation employs 50 or more persons, you are requred to have a written affirmative action program on file regardless of the size of your facility. Do you have a written affirmative action program on file?

   Yes _____ No _____

Figure 5.1 Notification to subcontractors, vendors, and suppliers.

5. Has your facility undergone a compliance review?

   Yes _____   No _____

   (a) By what agency?_____
   (b) Date of review   _____

_____   _____
          Name of Company   Location

CONTRACTOR'S

PROTECTIVE TERMINOLOGY FOR PURCHASE ORDERS

GENERAL CONDITIONS

All material and equipment furnished under this order shall be guaranteed by the seller to the purchaser and owner to be fit and sufficient for the purpose intended, and that they are merchantable, of good material and workmanship, and free from defects. Seller agrees to replace without charge to purchaser or owner said material and equipment, or to remedy any defects latent or patent, not caused by ordinary wear and tear or improper use or maintenance, which may develop within 1 year from date of acceptance by the owner or within the guarantee period set forth in applicable plans and specifications, whichever is longer. The warranties herein are in addition to those implied by law.

The seller of all material and equipment furnished under this order shall be subject to the approval of the architect, engineer, or purchaser, and the seller shall furnish the required number of submittal data or samples for said approval. In the event that approval is not obtained, the order may be canceled by the purchaser with no liability on the part of the purchaser.

All material and equipment furnished hereunder shall be in strict compliance with plans, specifications, and general conditions applicable to the contract of the purchaser with the owner or with another contractor, and the seller shall be bound thereby in the performance of this contract.

The materials and equipment covered by this order, whether in a deliverable state or otherwise, shall remain the property of the seller until delivered to a designated site and actually received by the purchaser, and any damage to the material and equipment or loss of any kind occasioned in transit shall be borne by the seller, notwithstanding the manner in which the goods are shipped or who pays the freight or other transportation costs.

The seller hereby agrees to indemnify and save harmless the purchaser from and against all claims, liability, loss, damage, or expense, including attorneys' fees by reason of any actual or alleged infringement of letters patent or any litigation based thereon covering any article purchased hereunder.

Time is of the essence of this contract. Should the supplier for any reason fail to make deliveries as required hereunder to the satisfaction of the purchaser, or if the materials are not satisfactory to the architect or engineer, the purchaser shall be at liberty to purchase the materials elsewhere, and any excess in cost of same over the price herein provided shall be chargeable to and paid by the supplier on demand. Should any delay on the part of the supplier, or should any defects or nonconformance of the materials or equipment with the plans and specifications occasion loss, damage, or expense, including consequential damages to the owner and the purchaser, the supplier shall indemnify the owner and the purchaser against such loss, damage, or expense, including attorney's fees. If, for any cause, all or any portion of the materials to be furnished are not delivered at the time or times herein specified, the purchaser may, at his or her option, cancel this order with respect to all or any portion of materials not so delivered.

The seller shall furnish all necessary lien waivers, affidavits, or other documents required to keep the owner's premises free from liens or claims for liens, arising out of the furnishing of the material or equipment herein, as payments are made from time to time under this order.

All prior representations, conversations, or preliminary negotiations shall be deemed to be merged in this order, and no changes will be considered or approved unless this order is modified by an authorized representative of the purchaser in writing.

Figure 5.1 *(Continued)*

6

Market Sources
for Revenue

The 8A program: A Source of Revenue for Disadvantaged and Minority-Group Contractors

Public Law 95-507, Section 8A, was enacted in 1969-1970 largely because of social unrest and a lack of jobs for people from minority groups. It is administered by the Small Business Administration (SBA), as of this writing, which acts as the prime contractor for government agencies in granting contracts and subcontracts to members of minority groups and handicapped people. To apply, a minority-group or disadvantaged business owner writes to the SBA and requests information about the program. He or she must prove disadvantage and historical discrimination. As of 1986 a business is allowed to enter the program for a fixed participation period: 4 years with an option to renew for 1 or 2 years. The acceptance process can take up to 6 months or a year. Because of the fixed participation period, you should begin marketing your services to federal government agencies immediately after acceptance, in order to take best advantage of the opportunity.

Most of my 8A experience was with the U.S. Navy, and they're tops! To get started, you might pay a visit to Naval headquarters in your geographical area and say you're in the 8A program and would like to know what type of work they have coming down the pipeline. For thi information you need to see the contract specialists. Another approach is to obtain an agency list from the Small Business Information Office (SBIO), which will tell you where to begin marketing your services. Or you may go to the planning and engineering department of the Navy or any other federal agency first, or to the SBA representative for that agency, and ask them what work is coming down their pipeline. This will serve as a checkpoint. Some of the contract specialists will fail to mention some information; and later on, if you keep up with the *Dodge Reports* and other informational sources, you may find you were misinformed. This is the system's way of controlling what you get. I've never understood this, because if you added up all the contracts minority and disadvantaged firms get from the federal government, they wouldn't even register on the calculator or computer when compared to federal contracts issued in general. (See the section below entitled "How to Win 8A Contracts." Large organizations with sales forces may also find the information in Figures 6.1 to 6.4 and Table 6.1 useful.)

TABLE 6.1   How Long is a Person in Your Organization Considered a Trainee?

| | Percent of Firms Reporting | | | |
	Consumer	Industrial	Service	All
Don't hire trainees	4.8	1.6	3.4	3.1
Variable	1.4	3.2	1.1	2.1
1–3 months	8.2	5.9	14.8	8.5
4–5 months	3.4	2.1	3.4	2.8
6–9 months	25.8	17.0	21.6	21.1
10-11 months	—	3.7	1.1	1.9
1 year	32.6	28.7	20.5	28.4
1.5 years	4.8	8.0	4.5	6.1
2 years or more	19.0	29.8	29.6	26.0
	100.0	100.0	100.0	100.0
Overall average number of months a person is considered a trainee	13.1	15.8	14.6	14.6

Qualifying for the 8A program

Although many minority people have been able to take advantage of the 8A program, the term "disadvantaged" can also apply to white people, under certain conditions, and to persons of other races who can prove that they are disadvantaged or handicapped, or have been historically discriminated against.

When the system was originally set up, not many skilled minority business enterprises (MBEs) or disadvantaged business enterprises (DBEs) were available. It is important to understand that, as soon as minorities and the disadvantaged became skillful enough to complete jobs and to obtain multimillion dollar contracts, and overcame some bonding handicaps, the agency began strictly enforcing Fixed Program Participation Termination (FPPT).

The SBA bonding program is a tool that has really helped MBEs and DBEs. Without "big brother's" bonding assistance, a lot of jobs would have been impossible for MBEs and DBEs to obtain. MBEs and DBEs functioning in this program should be grateful to "big brother."

SALES PLANNING AND POLICY

1. Estimates company's share of industry volume; sets long-range marketing objectives.
2. Informs top management of adverse or favorable competitive conditions, special marketing needs, and opportunities.
3. Develops long-range marketing programs around current and projected products and markets.
4. Plans and coordinates short-term programs needed to operate manufacturing facilities advantageously and to meet set objectives.
5. Transmits plans and programs to subordinate marketing executives; assigns specific action and follows up.
6. Participates in top-level marketing and field sales meetings to develop full understanding and acceptance of policies and programs.

ORGANIZATION, TRAINING, AND COMPENSATION

1. Determines organization needed to attain company objectives; directs its development and maintenance.
2. Appraises immediate subordinates and helps them to improve; directs them in appraising and improving their own subordinates.
3. Reviews recommended appointments to key marketing positions; appraises candidates' growth and potential; directs corrective action.
4. Directs the director of selling operations in development and use of recruitment, training, and appraisal methods.
5. Supervises development of salary and incentive schedules, and methods of crediting sales.
6. Directs administration of compensation plans.
7. Investigates organization and methods of employment, training, and compensation used by comparable outside concerns.

FORECASTS, QUOTAS, BUDGETS, AND CONTROLS

1. Directs gathering of information for and preparation of periodic sales forecasts.
2. Reviews proposed sales quotas; supervises setting of final quotas for all products, markets, and sales territories.
3. Reviews departmental budgets; directs consolidation into division budget; recommends desirable expenditures not included in budget.
4. Supervises allocation of approved division expenses to departments.
5. Consults with the controller's division on development of effective controls, forms, and report frequencies for sales and expenses; directs maintenance and analysis of records supplied by the controller's division; reviews these records regularly and directs corrective action.
6. Advises manufacturing division on changes and trends in sales and marketing plans that affect production scheduling and inventory policies.

SELLING OPERATIONS

1. Develops subordinates' understanding and acceptance of company and division objectives, policies, and programs.
2. Directs domestic and export selling and sales service operations; reviews operations and directs corrective action.

Figure 6.1 Sales planning and policy.

3. Directs coordination of all line and staff marketing operations, and development of team effort.

INDUSTRY AND CUSTOMER RELATIONS

1. Directs maintenance of good customer relations; determines needs and opportunities for improving service functions; directs corrective action.

2. Directs handling of emergency aspects of complaints about product quality.

3. Supervises percentage allocations of restricted merchandise between domestic and export selling departments.

4. Directs participation in trade association activities and important conventions; designates individuals for membership in such associations and attendance of conventions.

PRICING

1. Recommends pricing policies; establishes prices, terms, and discounts in line with company objectives; reviews proposed products and establishes their prices.

2. Studies competitors' prices, terms, and discounts.

3. Determines desirability of special prices, terms, and discounts to move slow inventories or to support special sales efforts.

4. Consults with company legal advisers to ensure that contract forms conform to applicable laws and regulations, are consistent with company policy, and are fair to customers and company.

5. Directs preparation of price lists for publication.

ADVERTISING, SALES PROMOTION, PUBLICITY, AND PUBLIC RELATIONS

1. Reviews requested advertising, sales promotion, publicity, and public relations appropriations and their ratios in relation to company objectives.

2. Directs coordination of advertising, promotion, publicity, and public relations programs with selling activities; directs development of understanding and effective use of these programs by selling departments.

3. Directs selection of and liaison with outside advertising, publicity, and public relations agencies.

4. Directs preparation of catalogs, bulletins, manuals, and point-of-sales materials.

5. Participates in professional and other groups, to the extent necessary and economically feasible, in order to improve and maintain the company name and recognition in industry and with the public.

PRODUCT AND MARKET RESEARCH AND DEVELOPMENT

1. Initiates and directs establishment of programs for obtaining information on product and market needs and opportunities.

2. Supervises analysis of information obtained; reviews it and directs appropriate action; submits requests for action outside the division.

3. Directs development of new-product specifications and proposed prices, packaging, packing, and marketing methods, as well as display, advertising, promotion, and selling ideas.

4. Supervises contacts between marketing division and outside design, laboratory, and testing services.

Figure 6.1 *(Continued)*

SALES OPERATIONS RESEARCH

1. Initiates and directs planning and conduct of marketing operations research.

2. Reviews information and suggestions submitted; determines corrective action.

3. Recommends needed changes and improvements affecting other divisions.

EMPLOYEE RELATIONS

1. Applies basic company employee relations policies in all dealings with subordinates; directs them in dealings with their subordinates.

2. Collaborates with employee relations division in developing understanding and acceptance of policies throughout the division.

3. Actively encourages development of a spirit of harmony, cooperation, and interest in company's welfare.

Figure 6.1 *(Continued)*

One advantage to securing work under the 8A program is that there is more potential profit in any negotiated job than in bid work. For example, if you could minimize risk while increasing volume and decreasing capital outlay, wouldn't you do it? Any sound-minded business person would. But what you have to understand is that in the 8A program, you must also be doing other work in addition to 8A.

New 8A program participants, historically, have bonding problems and are undercapitalized. You must initially maintain a 50:50 (or preferably a 60:40) ratio of 8A to other business, with immediate or gradual replacement of the lost SBA portion, upon FPPT. When consulting with 8A contractors, I work toward achieving a ratio of 80 percent commercial jobs to 20 percent 8A government work by the time FPPT goes into effect, in order to have a sound business volume.

Some businesses almost fail before they can get an 8A contract. For example, one 8A firm was excited about a large contract obtained via the 8A program and eager to do a good job. The business was small and had been in the program approximately 16 months. The firm was heavily dependent upon this contract, and it consumed all the firm's time. The firm was undercapitalized and couldn't hire anyone to continue the regular open-market efforts; and the owner couldn't do it either. After the firm had been working on the 8A contract for 10 months, the results were a cashflow bind, loss of revenue, and a second mortgage on the owner's home in a vain attempt to stay afloat. Finally, the firm went out of business.

I am not an expert on insurance or bonding, but I do know that these are two areas that need constant attention. Most government jobs, except service contracts, totaling $25,000 and up require bonds.[1]

---

[1] If you think you may fall within the government's definition of "minority," and you need assistance to obtain a bond, call me at 703-362-2875.

Advantages. Advocates of commission plans point to these advantages over other methods:

1. There is a high degree of incentive for sales representatives to get orders. Since the amount of their pay is geared directly to their selling efforts, the harder and more successfully they work, the greater their earnings. Sales representatives will often work longer and harder under a commission plan than under any other plan.

2. The risk to the company of incurring high sales expenses is minimized. Regardless of how sales representatives choose to operate, the company knows that its sales expenses cannot exceed the commission rate. A company wanting to open a new territory can do so without being concerned about the sales cost, for the risk is shifted to the sales representives.

3. The margin of profit on new business can be determined in advance.

4. Many good sales representative prefer to be paid by commission, which gives them a feeling of independence and opportunity that they believe they would sacrifice under another type of plan.

5. The plan's simplicity of operation appeals to many companies and many sales representatives. The computations necessary in more complex incentive plans are avoided, and problems of revising salaries are also overcome. The commission plan is flexible in good times and bad; the company pays well when it can afford to and economizes automatically when buying is restricted.

6. If commission rates are equitable, sales representatives know that their pay is not based on someone else's judgement of their value, and that there is no question of unfairness in comparison with other sales representatives.

Disadvantages. Disadvantages of compensation by the commission plan are important and numerous. Some of them can be reduced considerable by an awareness of their existence and a tightening of supervisory control. Others explain the trend away from straight commissions in favor of combination plans; for example:

1. Pressure for sales volume dominates the commission sales representative's thinking. It encourages overselling. Whereas the long-term view would dictate against overstocking a customer, the short-term desire for bigger commissions may make the sales representative adopt unsound selling practices.

2. Lack of attention to balanced selling is a corollary of the drive for volume. If a particular part of the line sells readily, the sales representative may push it and neglect other items. Similarly, they are not so much concerned with building profitable accounts over a period of years as with aggressive selling to any accounts that will buy from them.

3. A sales representative is encouraged to skim the market. They wish to pick up the quickest and easiest possible sales. Companies with commission representatives may have relatively thin coverage of their markets, and yet the representatives may be clamoring for larger territories. Thus, where a company has an established market and is attempting to cultivate it intensively by building up goodwill and repeat business, a commission plan may prove disadvantageous.

4. Commission sales representatives are hard to control. They consider that they are essentially in business for themselves and that their methods are their own business, since their methods determine their pay. Where a company has indirect selling jobs to do, where there are junior sales representatives to be assisted in the field, or where reports require an appreciable amount of the sales representatives' time, there may be poor response under the commission plan.

5. Cooperation between sales representatives may be poor. Field sales effectiveness may call for assistance by one representative in covering another representatives' territory, or joint efforts in covering certain customers. Commission sales representatives want total credit for all customers and no interference from their associates. Split commissions are invariably an headache.

6. Earnings are subject to wide fluctuations. In boom times, commission sales representatives may obtain large earnings with a minimum of effort. Conversely, their morale may

Figure 6.2 Advantages and disadvantages of commission plans.

suffer in slack periods because of inadequate incomes even with hard work. The commission sales representatives' pay is not related to the relative amounts of effort in different stages of the business cycle.

7. A new sales representative finds an adequate income difficult to attain under a commission plan.

DRAWING ACCOUNTS. The use of a drawing account as a modification of the straight commission plan does not change the applicablily of the plan in any important respect. The drawing account consists of a weekly or monthly amount of money advanced. It is normally assumed that earnings from commissions will exceed amounts drawn, and thus these payments are advances only. Unless a drawing account is guaranteed (in which event it becomes a salary), the principal benefit to sales representatives is to allow them a more even spread of earnings throughout the year and during periods of sickness and emergency.

Some companies operate their drawing accounts on the principle that sales representatives will draw regularly throughout the year; only at the close of the season will they be permitted to draw the balance, up to the total amount earned by commissions. In other instances, representatives receive commissions frequently, and the drawing account represents only a safety factor.

One of the dangers of a drawing account is that it can stablize representatives' income over only a relatively short time. If difficulties last for a longer period, representatives may soon overdraw. Then they lose incentive; they see all their efforts being directed toward reducing an old indebtedness instead of providing current income.

Rather than continue under this burden, a sales representative will frequently resign when the drawing account is substantially overdrawn, and make a fresh start with another company. To guard against losing money in this way, some companies set the drawing accounts so that total funds available during the year are a stipulated fraction of expected earnings. It is thus unlikely that representatives will close the season with a balance owing to the company. Often, however, this is not conducive to good morale and company loyalty.

The frequency with which the drawing-account balance is closed against the commission credits depends largely on the amount of seasonal sales variation. If seasonal influences are small, commissions can be paid every month. A company with a strong spring and fall business may find it desirable to close the drawing accounts twice a year. Since representatives build up their credits early in the accounting period and then draw on them during the slack period, the company is protected against losing a representative with a large balance owing.

DETERMINING COMMISSION RATE. Implied in a flat commission rate is the proposition that all business is equally profitable to the company regardless of source or what is sold. Thus, it would seem desirable to use a flat rate only if the product line is relatively homogeneous and the marketplace is relatively unsegmented. The actual amount of rate will, of course, be determined from the standpoint of cost-revenue relationships. Also, as with any mode of compensation, the rate has to be competitive in the marketplace if appropriate personnel are to be attracted and held in the sales force.

In terms of breakeven thinking, there is much to be said for a progressive rate, inasmuch as the profit to the company should increase as volume increases. Arguments against a progressive rate center mainly on the possibilty of windfall income. A regressive rate runs counter to cost-revenue relationships but is used chiefly as a guard against windfall.

Another consideration that management must face in setting rates is whether there is to be a special incentive built in for the generation of new accounts as distingushed from continued sales to existing accounts. Ideally, if there are factors that dictate the use of a commission plan, the base on which it is computed would be profit, not gross revenue. Probably also progressive rates should be established differently across the product line and across the segments of the marketplace. If progressive rates are to be implemented, the company faces some very difficult policy decisions; perhaps the most critical one is whether to share profit figures with the people in the field. Furthermore, some companies are not in a position to compute a profit in refined enough fashion to isolate it by account categories and by products. Since rate of commission is determined by specific sales problems and cost-revenue relationships in each company, comparing actual rates from company to company even in the same industry, much less across industry lines, is relatively meaningless.

Figure 6.2 *(Continued)*

MAJOR TYPES. The possibilities with combination plans are even more numerous than with commission plans. The important categories are described below. In general, the salary ingredient should provide the sales representative with sufficient income to cover basic living costs, with the commission, bonus, or both as the extra incentive to achieve maximum results in all phases of the job.

Salary plus Commission. Variations include companies that pay a salary plus Commission on all sales, salary plus commission on sales exceeding a stated amount of total revenue, salary plus a differential commission on items in the product line, salary plus a differential commission on various account categories, and salary plus a differential commission on old and new business. In addition, the rate may be set progressively, regressively, or flat.

Salary plus Bonus. Companies encountering difficulty in isolating the impact of individual representatives' efforts often resort to this plan. Variations include bonus based on both qualitative and quantitative judgments about the representatives' productivity, bonus based on "profit pools" within sales districts, and bonus based on overall profitability of the enterprise. Usually the salary segment is paid either semimonthly or monthly, and the bonus monthly, quarterly, or annually.

Salary plus Commission plus Bonus. Companies using such plans attempt to provide incentive for actual selling efforts and results with commission, but in addition provide an incentive bonus for the successful performance of many collateral duties such as merchandising, training dealer sales representatives, and the like. The bonus segment is also subject to many variations. It may be an individual matter, it may be a share from a district bonus pool, or it may be based on the overall profitability of the enterprise.

VERSATILITY OF COMBINATION PLANS. There is no doubt that combination plans provide maximum flexibility and that today they are the most popular approach to compensation. They can be geared to a wide range of different requirements and selling conditions. Their greatest flexibility, of course, is in the commission. Here are some of the possibilities:

1. Commission on all sales (single rate).
2. Commission on all sales (single rate up to quota; second rate over quota).
3. Commission on all sales (single rate up to quota; increasing rates by specified amounts over quota).
4. Commission on all sales (single rate up to quota; decreasing rates by specified amounts over quota). (This is a weak type of plan.)
5. Commission on sales above quota (single rate).
6. Commission on sales above quota (multiple rates increasing by specified amounts over quota).
7. Commission on sales above quota (multiple rates decreasing by specified amounts over quota). (This is also a weak plan.)

Each of these possibilities, of course, can in turn be varied by differential rates for single products or product classes and by account categories.

ADVANTAGES. Proponents of combination plans suggest that these plans

1. Provide sales representatives with the sense of security that comes from a fixed income.
2. Provide a direct financial incentive to sales representatives and make compensation more commensurate with results obtained than under the straight salary plan.
3. Provide management with control over sales representatives, allowing for balanced selling efforts by products, customers, territory coverage, and so on.
4. Are more flexible than the straight salary plan, and better able to adapt quickly to changing conditions.

Figure 6.3 Advantages and disadvantages of combination plans.

---

5. Encourage missionary and merchandising activities.

6. Encourage sales representatives to be loyal to the company.

7. Encourage more cooperative effort with other sales representatives.

8. Make the company share financial risk with sales representatives.

9. Encourage sales representatives to push items on which the greatest company profit is made.

10. Assure selling costs in line with the volume of business secured.

11. Maintain sales representatives', morale through fairness, simplicity, and adequacy of reward for maximum effort.

12. Minimize disagreements relating to salary increases.

DISADVANTAGES.   There are, however, some disadvantages to combination plans, which must be considered:

1. Sales representatives may place too much selling emphasis on immediate volume if the commissions are large.

2. Sales representatives are less amenable to sales control when the bulk of their compensation comes from commission.

3. Combination plans can be complex and difficult to understand, particularly when a number of commission rates are applied to a long line of products. Such complicated plans also increase accounting costs.

4. Representatives' paychecks may be held up pending calculation of commissions.

5. In order to provide a suitable salary and reasonable selling costs, commission percentages may be too low to provide adequate incentive.

6. Such a plan may not offer adequate incentive to perform special sales jobs, nonselling tasks, or territorial promotional and development work.

Figure 6.3   *(Continued)*

Studying Your Employees and Associates

In my consulting business I plan the growth pattern for electrical contractors, establish their financial ratios, and attempt to help them understand others' behavior. It is important that, as you grow, someone in your organization (if not you yourself) study your employees and associates, and try to figure out what makes them tick or what their highest priorities are. Then you must also become sharp enough to pick up any changes in their needs and values. Remember, what turns a person on today may turn that same person off next year—and unnoticed changes could be devastating to your business. To have a good "people base," you must create a nucleus. For example, you could choose five good, loyal, dedicated, hard-working people as key field and administrative employees, and have each of them choose five, and so on. The only thing

The examples of the salary-plus-commission plan that follow represent a typical cross section of plans in use. They do not necessarily meet all of the requirements for a sound plan.

STEEL PRODUCTS COMPANY. A manufacturer of steel and steel products pays sales representatives a base salary to meet living-cost esssentials. In addition, commissions are paid on every dollar of shipments. Dollar value of shipments is calculated on the basis of classification of products into A, B, and C classes. Commission limits are established on business from any one customer for any type of product in a single fiscal year. In order to equalize sales representatives' opportunities, limits are set on territory potentials.

WOMEN'S APPAREL COMPANY. A women's apparel manufacturer selling through selected accounts pays its sales representatives a base salary that averages approximately two-thirds of their total compensation. In addition, they are paid increasing commission rates on sales over a quota, with rates increasing at specified amounts over quota.

WOMEN'S ACCESSORIES COMPANY. A women's accessories manufacturer selling through selected dealers and requiring considerable merchandising work pays salaries amounting to approximately 75 percent of total compensation. An upward-sliding scale of commissions is paid on all sales within specified dollar volume brackets, but there is no quota.

FOOD-PROCESSING MACHINERY COMPANY. A manufacturer of food-processing machinery selling through distributors pays salaries to its sales representatives according to territory classifications of A, B, and C. Classification of territories depends on

1. Potential or actual importance of territory from a sales-volume standpoint

2. Relative costs of living

3. Type and caliber of sales representatives required to serve the territory effectively

Each representative is paid a commission on all sales from his or her territory. Commissions are varied by classes of products. Provision against windfalls is stating that no commission will be paid on sales to one customer above a specified amount during a specified period.

BUILDING-MATERIALS COMPANY. A building-materials manufacturer gives each sales representative a quota, and his or her salary is a fixed percentage of this quota. Quotas are set annually, but there is a minimum salary limit. Commissions are paid on all sales. The various products are grouped into four classes. For three of these product classes, rates of commission up to and over the assigned quota are the same. For the fourth product class, one commission rate is paid on all sales up to the quota, and progressively increased commission rates are paid according to specified amounts over the quota.
No commissions are paid on

1. Goods sold below regular price

2. Sales in which sales representatives assisted in pooling operations

3. Replacement or donated materials

4. Bad-credit sales

BONUS PLANS

DEFINITION. Tosdal* has described this type of plan as follows:

The bonus differs from the commission in that there is no necessary mathematical relationship between performance and the bonus. The bonus is a lump-sum payment which may or may not be graduated in accord with the performance goal set up, but it is always awarded for better-than-average performance. A bonus which varies directly with performance base on the uniform, progressive, or regressive rate is not a "bonus" in the true meaning of the term.

*Tosdal, *Salesmen's Compensation*, vol. I.

Figure 6.4 Typical examples of salary-plus-commission plans.

The bonus element in a compensation plan provides an incentive for sales representatives to accomplish specific sales tasks. It is used in conjunction with a salary and sometimes in addition to a salary and a commission. Commission and bonus are rarely used in combination, although in some instances this is a feasible type of plan.

Basis for Bonus Payments. The bonus may be paid in accordance with how successfully a sales representative

1. Attains quota
2. Sells a balanced line
3. Increases total volume
4. Expands sales of more profitable products
5. Secures maximum coverage
6. Develops new accounts
7. Increases the number of calls made
8. Secures retail store displays
9. Performs desirable missionary work
10. Uses special promotional material
11. Does sales service work
12. Makes demonstrations
13. Develops new sales and product ideas
14. Follows up on sales development work
15. Performs other field activities that are important to the company's success

ADVANTAGES. The principal advantages of the bonus type of plan are that it

1. Most closely meshes the financial incentive with the specific sales activities which experience shows make for maximum sales success, thus providing an incentive for any or all of the elements of the required sales job that can be measured or appraised
2. Stimulates a balanced sales job
3. Is entirely flexible, because the incentive emphasis can be shifted to meet changing conditions and selling objectives without changing the basic plan
4. Compensates sales representatives most closely in relation to how successfully they perform the specified sales job
5. Permits direct control over selling activites and good control over selling costs
6. Can provide for sales representatives the security of a salary, plus an incentive for performing other sales activities that build and strengthen the company's position in the territory

DISADVANTAGES. The bonus type of plan has fewer disadvantages than any other basic type, However, it may be

1. More complicated because there is a tendency to try to do too much through a bonus plan
2. More difficult to operate since accounting and record keeping usually have to be somewhat greater than under straight salary or commission plans
3. More demanding of sales management inasmuch as success depends in part on a full and continuing knowledge by management of market requirements, buying habits, field selling conditions, how sales representatives should operate and are actually operating, and the like.

Figure 6.4 *(Continued)*

that never changes is that nothing stays the same, particularly when it comes to people.

How to Win 8A Contracts

Introduction

If you follow the information in this marketing approach to the letter, you should get good results. The time-tested pointers given below are based on my own firsthand experiences during many years of successfully managing people and marketing 8A contracts. As a former 8A contractor now affiliated with RACO Publications, I am sharing with you the techniques I used to build a very successful enterprise. This extremely helpful information can help you to gain rich rewards for your procurement efforts.

Marketing Plan for the Procurement of 8A Contracts

The 8A program has much to recommend it, but there are some deterrents which you may encounter while seeking out contracts. The history of direct and indirect participation in this field has provided specialized knowledge which has helped in this writing.

Many changes are taking place in the 8A program, caused by lobbyists from companies that oppose the 8A program, even as I write this book. Remember that you must always obtain up-to-date information on the current status and provisions of the 8A law from the SBA.

Below are some timely tips on steps you should take before you actually begin marketing:

- After receiving certification in the 8A program, make firm and workable plans for properly maintaining the sales that supported your business before you entered the 8A program. If you do this conscientiously, your eventual exit from the 8A program will go more smoothly.

- Remember that the best marketing effort requires that you do your homework very well. This homework includes identifying and studying the geographical area in which you plan to work. You will also need a business card and a descriptive brochure, if you don't already have them. Your business card should indicate your company name, street address or post office box, the principal officer or officers, your business telephone number (including area code), the type or types of work you do, and the number of years you have been in business, especially if over 2 years. Your brochure should be detailed, showing the name, location, telephone number, contact person, and dollar value of previous jobs. Also include copies of letters of commendation received upon satisfactory completion of jobs (get *writer's* permission), bonding capacity, and your agent's name and telephone number. These are vital selling tools. Don't overlook any of them.

Let's look at a potential problem area. You are new at seeking 8A work, and you will most likely lack the specialized knowledge required to make a successful presentation. Government agency representatives are aware of this lack and usually capitalize on it. The less informed you appear to be, the more "unaware" the agency representative may pretend to be. It is therefore in your best interests to learn as much as possible before making your initial presentation.

In many cases, an agency representative holds the title "contract specialist" or the position of department head. If you do not know who the agency representative is in the agency you plan to approach, you can contact the business development specialist (BDS) at your local Small Business Administration office and obtain a marketing list. Here are some important things to do before your appointment.

- Pay a visit to the planning and scheduling department of the agency in which you are interested.

- Find out the name of the planning and scheduling department head, arrange to meet that person, and try to establish favorable rapport. The following questions are good ones to ask:

1. "What do you have on the project board?"
2. "What is the status of the projects? Are they 60, 90, or 100 percent complete?"
3. "Have they been reviewed by the owner?"
4. "What is the approximate dollar value? I need to know for bonding purposes."

After obtaining the information you need, ask the department head for the project ID number. This is the magic key and should be considered and treated as public information, in my opinion. But you should be cautioned: The project ID number is indeed very difficult to get. However, you must try, because it is the key to the entire 8A marketing ball game.

The Appointment. Here is an example outlining an initial attempt at procurement of an 8A contract from an agency. The characters are Bob Jones, an employee of company X, and Jane Smith, an agency representative (AR). Bob Jones makes a telephone call to the agency with which he wants to do business. The conversation should go something like this:

BOB JONES: Hello, my name is Bob Jones. I'm with the X Company. My company is in the 8A program, and I would like an appointment to speak with Ms. Jane Smith about negotiating 8A contracts. My company specializes in building and remodeling.

After the agency representative comes on the line, the conversation continues.

AGENCY REPRESENTATIVE: Hello, Mr. Jones. I can see you on Monday, January 6, at 8:30 a.m.
BOB JONES: That's just fine. What building is your office in?
AGENCY REPRESENTATIVE: My office is in Building 25, Room 609.
BOB JONES: Good! I'll see you on the 6th.

You may, however, want to ask for a later hour for your appointment than the one mentioned above, in order to leave time for the important entrance procedures described below. It has been my policy to allow at least one extra hour to complete these procedures without feeling hurried.

Entrance Procedures on Day of Appointment.

- You must obtain visitor and automobile passes at the pass gate.
- At the pass office, you must present the following items: your picture, ID card, registration card and proof of automobile insurance, or a copy of a leasing agreement if you're driving a leased vehicle.
- At the pass office, you will also be asked the nature of your business and what person and building you will be visiting. Pass office personnel will check with the agency representative to confirm your scheduled visit.

In any high-security area, only after all these entry procedures have been satisfactorily performed will you be allowed to enter the main premises. Always remember that a polite manner and a warm smile will go a long way toward making your visit a pleasant, successful one. Speak cordially upon entering the working area of the agency representative's secretary, and do not enter the representative's office until the representative has been notified of your arrival and you have been told to go in.

Greet the agency representative with a genuine smile and good eye contact as you extend your hand in a firm handshake. A little preliminary small talk (about the weather or the office furnishings, for example) will help to lighten the atmosphere before you bring up the reason for your visit. The following example could be helpful.

BOB JONES: Good morning, Ms. Smith. It's certainly a bright and lovely day.
AGENCY REPRESENTATIVE: It certainly is, Mr. Jones. It's a pleasure to meet you. Have a seat.
BOB JONES: Thank you for seeing me.
AGENCY REPRESENTATIVE: Well, what can I do for you today?

BOB JONES: I would like to go over your project list for the upcoming fiscal year, October 1 through September 31.

At this point, you may find out whether or not you're dealing with an agency representative who is genuinely interested in helping you obtain your company's 8A goals. An agency representative who is working in your interest will share helpful information with you, including showing you the project list and ID numbers or, at least, reading them to you.

The way you handle yourself during the initial meeting will show the agency representative whether you've done your homework. If the agency representative seems reluctant to reveal the project list and ID numbers, it's a good idea to refer to other information (including ID numbers) which you have previously obtained from other contacts. After this revelation, you should again inquire about the project list and ID numbers—but be prepared for some possible slight resentment. If you follow this procedure, you should become rather shrewd at obtaining 8A contracts.

Besides installation contracts, also ask about maintenance contracts. These are attractive because they do not require a bond.

If you do receive the information you need from the agency representative, including the ID numbers of upcoming projects, you will have the necessary sales data to make a telephone call[2] to the business development specialist at the local Small Business Administration office. Give the business development specialist the ID number of the job you are interested in, and request that he or she either notify you or see to it that you get this job when it comes through the SBA. Active involvement will let the business development specialist know that you're making a good effort to stay on top of things, rather than just sitting back waiting for the SBA to do your job for you.

Another plus would be for you to request that the agency representative ask that your company be given the job when routing the project through the Small Business Administration. This method has worked well in many areas of the country, although Washington, D.C., has been one exception. In the D.C. area, the marketing focus is seemingly aimed at the SBA, with moderate attention given to the local agency.

After finding out from the agency what projects have been sent to the SBA, seek verification from the SBA and make your inquiry about possible procurement. Over the years, the Richmond, Virginia, Small Business Administration has had the best nationwide system for procurement through Section 8A, and has received high recognition for its excellent record.

Again, a caution: Do not totally depend on the 8A program. Remember that it is controlled by people who could have a rather decisive

---

[2]As of this writing, some SBA offices are requiring that the request for projects be made in writing.

hand in your financial destiny. Make every effort to be both competitive and business-wise in the open market. Make capitalism work and work well for you.

The 8A program, when properly used, is an extremely valuable way for the qualified firm to work toward financial viability. I wish you the very best luck in your work with 8A contracts. They have worked extremely well for many companies. They can work as well or better for your company.

Labor Reporting and Cost Recovery

Proper reporting of labor, material, and direct job expense (DJE) is critical. Every estimate should be categorized by labor, by materials, and by DJE. This is actually where the job budget comes from. Changes in cost must be reported daily, at the end of the day. The field people should be told the number of hours for a given installation and the number of units that must be installed daily to meet the unit installation labor budget. By monitoring this information you can readily see whether production will be on target or not, and what adjustments must be made. In order to make money in the electrical contracting business, the following are musts:

1. The estimate must be accurate as to all cost recovery.

2. You must have competent personnel.

3. The right personnel, information, materials, and tools must be on the job site in a timely manner.

4. The work must be properly planned and scheduled, and must be installed within the estimated cost.

When these conditions are met, expecting a profit is realistic. When these conditions are not met, don't blame the people in production; blame top management. The fact is, however, managers are no better than the people they are surrounded by. If you're surrounded by incompetent people, fire them! Otherwise, you'll pay them to break you.

There is a secret to getting people to do what you want them to do. The secret is appealing to their needs. The "catch 22" is the necessity for keeping pace as people's needs and values change, as mentioned before.

Field Labor

Daily log reporting is essential and could very well represent money in the bank for you, especially on government jobs. (This is not to say, however, that you should take log reporting on other jobs lightly.) If you are a newcomer to government contracts, I can assure you, you'll lose money if you fail to document or if you take it upon yourself to delete or install items not shown on the plans or included in the contract documents. You must get authorization in writing for all charges.

When to Subcontract

Although electrical contractors are usually subcontractors, there are times when they can subcontract their work, or a portion of it, and make more money than if they did it themselves. In this situation the "catch 22" is: Always be certain that the contract documents allow you to subcontract.[3] You should subcontract when:

1. You need added work force.
2. You discover you will make more money if you subcontract.
3. You don't have the expertise for a certain portion of your contract.
4. Financing will cost you less if you subcontract.
5. You don't have the necessary equipment.

Non-8A Markets

1. *Private-sector projects.*   Competitive bids in the private sector usually take care of themselves with little or no problem, and with little or no profit. No one cares how much you make if you're low bidder; you will not be hindered by a string of public laws stretching from Maine to Florida about defective pricing and return of so-called excess profits.

2. *Municipal and state jobs.*   These jobs are full of "loopholes," and at the same time, stringent requirements apply. These contracts are very similar to federal contracts.

3. *Federal government work.*   Among the fairest when it comes to items such as different site conditions and specifications in general are federal government jobs. The bidding is the fairest. All you have to do is show up on time at the bid opening and be the low bidder—and you've got the job, provided you meet the other requirements. There's no price shopping and the checks are good. There could, however, be some excess time lapse in receiving your money, though I personally have not experienced this. In performing government work, keep in mind that you cannot make excess profits. The federal government gives itself up to 3 years to audit your records. "Big brother"

---

[3] Note—Be certain that you bind the subcontractor to you and to the owner the same as you are bound, directly or indirectly. Use an American Institute of Architects (AIA) standard subcontract form.

has a lot of bureaucratic procedures for you to go through. If you're new, you'll need help in certain areas of your paperwork and in understanding how the system works. To prevent defective pricing, you must not have obtained a lower price than the one you turned in as of the date you signed the truth-in-pricing form, but a lower price dated after the date on the truth-in-pricing form is permitted. You absolutely must keep good faith on government projects. Big brother does not play games.

Education

Education is an ongoing, never-ending process. In order to take advantage of all possible market sources of revenue, you must attend seminars, workshops, and trade shows; take courses at community colleges; and also explore any other educational opportunities that may be available to you.

Promoting Your Business

Success and time management go hand in hand. A person who cannot manage time wisely cannot achieve maximum success. Promoting your business demands a multitude of activities, most of which must proceed simultaneously. These include

- Developing your image
- Familiarizing yourself with media outlets
- Developing and maintaining a self-promotion network
- Locating and using sources of information
- Creating and producing articles, press releases, etc., about yourself and your company

Self-Promotion

As Americans, we are always in search of something better. First-class self-promoters link themselves with a powerful image through the process of defining their work and their accomplishments as they relate to the free-enterprise system. Adept self-promoters are dedicated to the mission of enriching others and society in general—not just themselves. For the most part in the past, the concept of entrepreneurial image was applied primarily to men. This is no longer true. As a shrewd self-promoter, a woman can create and make excellent use of an entrepreneurial image.

As a self-promoter, you must understand that, although most people basically mean well, they are usually slow to deliver. This means that if you want something to happen, you must go out and make it happen.

SAMPLE ELECTRIC COMPANY
ELECTRICAL CONTRACTOR

2400 Sample Avenue
Sample, OH 23760

"QUALITY & CRAFSMANSHIP
SINCE 1960"

- Class A Contractor
- Master Electricians
- Insured & Bonded
- High-Voltage Skills
- High-Voltage Skills
- Minority-Owned

CALL US TODAY:

(513) SAMPLEE

SAMPLE ELECTRIC COMPANY
ELECTRICAL CONTRACTOR
2400 Sample Avenue
Sample, OH 23760
— — — — —0— — — — —
"Quality & Chraftsmanship
since 1960"
— — — — —0— — — — —

Would You Like To Have
More Winning Bids?

Then please consider using our
services as an electrical subcontractor.

Twenty-five years of experience, a
reputation for doing the job on time,
and doing it right the first time, and
very cost-competitive operating
methods are just a few features that
you'll like about us.

A seasoned work force that can work
with high voltages (37.5-systems, for
example) just as easily it works with
normal electrical systems, is ready to
respond quickly and efficiently to your
requirements.

We are now embarking upon an
expansion of our resources. I'll be
calling you soon to see if it would be
beneficial for us to have an
introductory meeting. You'll see how
others have enjoyed the results of our
electrical contracting services!

Want more information? Then, without
obligation, call us right now at
(513) SAMPLEE!

Figure 7.1 An example of an attractive and effective company brochure.

SAMPLE ELECTRIC COMPANY, ELECTRICAL CONTRACTOR

Sample Avenue, Sample, OH 23760

— 0 —

"Quality & Craftsmanship since 1960"

Where Do We Work?
- Institutions
- Businesses
- Factories
- Industrial Plants
- Utilities
- Commercial buildings
- Apartments
- Marine Applications
- Government projects
- Schools, colleges
- Hospitals
- Churches
- Offices
- Residences & homes
- Service stations

What Kind of Projects?
- New construction
- Modifications
- Upgrading
- Rehabilitation
- Additions

What Kind of Applications?
- Substations
- High-voltage systems
- Power panels
- Circuit Breakers
- Standby power systems
- Energy-saving systems
- Heat pumps
- Heating
- Air conditioning
- Refrigeration

PHOTO OF EQUIPMENT, STAFF

We Are Ready to Serve You!

--- 0 ---

For Your Information:
- Association Memberships
- State Contractor Va., NC
- Graduate 8 A Program

--- 0 ---

Satisfaction Guaranteed

--- 0 ---

Call (513) SAMPLEE

--- 0 ---

What Kinds of Work?
- Power Wiring
- High-Voltage Services
- Cable Pulling and Splicing
- Electrical Repair
- Electrical Installation
- Substations
- High-Potential Testing
- Lighting, Wiring, & Fixtures
- Signalization
- Roadway Lighting
- Underground & Overhead Utilities
- Maintenance Contracts
- Trouble Shooting
- Design & Installation
- Service Changes
- Electrical Inspections
- Control Systems

What Kinds of Electricity?
- Alternating Current
- Direct Current
- Up To and Including 37,500 Volts

Why Not Call Us Now?
- Class A Contractor
- Master Electrician
- Insured and Bonded
- Inspection Service
- Free Quotes & Bids
- Established 1960

Figure 7.1 *(Continued)*

Printed Materials

Attractive, well-prepared printed materials are a must for the serious self-promoter. (See Figure 7.1.) Your printed advertising materials are a direct reflection of you and your business. They either add to or detract from your image as a skilled professional.

Logo

A professionally prepared logo is a wise investment if you can afford it. Shop around, by telephone, with printing companies and graphic design firms. Most will quote a range of fees over the phone. Try to find one that has experience with the construction industry.

Business Card and Letterhead

Your business card need not be an expensive, engraved one, but neither should it be of the cheapest quality. The same holds true for letterhead stationery and other materials. While it is always important to avoid unnecessary expenditures, you should avoid making yourself—and your business—a victim of the old "penny-wise, pound-foolish" syndrome.

Promotional Brochure

At all costs, avoid sending out handwritten brochures or flyers. In the minds of many people, such amateurish promotional materials signal you as an amateur electrical contractor. You must at least have your brochure neatly typed, and of course a printed brochure, if you can afford it, is preferable.

The Media

Media relations is not a subject just for those in "glamour" professions. There are many advantages to getting to know and becoming known by radio, television, and newspaper people. Whom do reporters call for expert information on a particular subject? Those persons whose names they know, of course. Anytime your name or your company's name appears in the media, it's publicity.

Develop your company's media relations by sending announcements and press releases on all newsworthy events involving your company. Here are some examples of appropriate occasions for sending out press releases:

- When you receive a noteworthy commercial contract (provided your contract does not prohibit publicity)

- When you or one of your employees is elected to office in a trade association

- When you or a representative of your company does something to benefit the community

Another good way to get publicity, while also performing a public service in educating the community, is to write letters to the editors of local publications expressing your opinions, based on your professional knowledge, about local issues affecting or relating to the construction industry.

Summary

The suggestions given here are just a few examples of the many opportunities you will have to promote yourself. Use all the opportunities you can. It's not enough just to be a good electrical contractor; you must spread the word about how good you are. Promote yourself by use of the following:

Press releases

Brochures

Business cards

Sales letters

Achievement of Goals

To achieve personal fulfullment, each person should set and reach the goals that will help to satisfy his or her personal need to excel. Each of us has many different needs. The first step in goal setting is to identify the areas in which you have the greatest need to excel. Those areas are what motivate you.

The following exercises are designed to help you gain some insight into the areas that are of greatest personal importance to you. The more adept people become at identifying and positively satisfying their own needs, the more motivated and mature they become.

A 5-Year Projection

Project yourself 5 years into the future. How old will you be in 5 years? What will you be like then? How will your personal, family, and career circumstances have changed by that date? Even realizing that this is a highly imaginative projection, you should attempt to be as realistic and objective as possible.

In working on the projection shown in Figure 8.1, you will probably be repeatedly bothered by two questions:

Should I describe my future the way I want it to be?

or

Should I describe my future the way I really think it is going to be?

You will probably allow both factors to enter into your answers. Such a solution is both natural and desirable. This projection is for

1. In 5 years my age will be _____ .

2. My occupation is (be as specific as possible)_____
   _____

3. My specific responsibilites are_____
   _____

4. My (or my household's) approximate annual income is_____
   _____

5. My most important personal possessions are_____
   _____
   _____

6. My family responsibilities in 5 years will be_____
   _____
   _____
   _____
   _____
   _____

7. Of my experiences in the last few years, the most pleasureable were_____
   _____
   _____
   _____
   _____
   _____
   _____
   _____

8. Of my experience in the last few years, the ones that gave me the greatest sense of accomplishment were_____
   _____
   _____
   _____
   _____
   _____
   _____
   _____
   _____

9. In the last few years, several dramatic things have happened in my business, community,
   or both, which gave interested me. Below is a summary of the highlights, including a description of how I was involved in these events.
   _____
   _____
   _____
   _____
   _____

Figure 8.1 Five-year projection.

_____
10. In reviewing my 5-year projection, the most important observations I made were_____

_____

Figure 8.1 *(Continued)*

your benefit. No one will see this projection other than yourself, unless you wish to share it with someone whom you feel close to.

"Who Am I?"

Now that you have imagined the future, let's return to the present by filling out the questionnaire in Figure 8.2. The purpose of these exercises is to aid you in drawing together information about yourself. You will use this information in goal setting and planning.

Crystallizing Your Thinking

Once you have identified some of the areas in which you need to excel, you should begin to crystallize your thinking. What goals can you set and achieve that will satisfy those needs?

Shown in Figures 8.3 and 8.4 are several exercises. They should help you to:

1. Open up your thinking. Too many times we prejudge that we can or can't do something before we allow ourselves to want it.

2. Make your goals tangible and measurable. Too many times we set only intangible goals that can't be measured, and as a result, we don't know whether we won or lost.

3. Become specific. The more specific we become, the easier a goal is to reach.

1. What aspect of what you are doing now gives you
   a. The greatest sense of accomplishment?_____
   _____
   _____
   _____
   _____
   _____
   _____

   b. The least sense of accomplishment?_____
   _____
   _____
   _____
   _____
   _____
   _____
   _____
   _____

2. What are your personal strengths?
   _____
   _____
   _____
   _____
   _____
   _____
   _____

3. In what areas do your personal strengths need development?
   _____
   _____
   _____
   _____
   _____
   _____
   _____
   _____

4. Many people dream of a "secret project" or a very private plan for "sometime" in life. What is your *real aim* in life?
   _____
   _____
   _____
   _____
   _____
   _____
   _____

5. Many of us would like to have the freedom to do the things we want to do when we want to do them. What would you do if you had:
   a. One hour_____
   _____

Figure 8.2   Present-day inventory.

_____
_____
_____

b. One day_____
_____
_____

c. One week_____
_____
_____

d. One year_____
_____
_____
_____
_____

6. Answer the question, "Who am I?"
_____
_____
_____
_____
_____
_____
_____
_____
_____
_____

7. Imagine that a statue of you is to be created. What would you want to list on the plaque as the major accomplishment of your life?
_____
_____
_____
_____
_____
_____
_____

8. As a result of reviewing what you have written, you feel that the areas that motivate you are:
   a. _____
   b. _____
   c. _____
   d. _____
   e. _____

Figure 8.2 *(Continued)*

f. _____

g. _____

h. _____

i. _____

Figure 8.2 *(Continued)*

List below everything you've ever wanted. Put check marks by the six most important.

_____
_____
_____
_____
_____
_____
_____
_____
_____
_____
_____
_____
_____

List below your six most important goals. Describe each as specifically as possible.

1. _____

2. _____

3. _____

4. _____

5. _____

6. _____

Figure 8.3 Inventory of goals.

List below your six most important goals. Describe each as specifically as possible.

1. _____
   _____
   _____

2. _____
   _____
   _____

3. _____
   _____
   _____

4. _____
   _____
   _____

5. _____
   _____
   _____

6. _____
   _____
   _____

Figure 8.4 Specific goals.

4. Deal with conflicts. Conflicts occur when we want to do two things at the same time—and we have an equal amount of desire to do both.

Converting an Intangible Need to Excel into Tangible Goals

If you have a large number of intangible needs, or one very strong intangible need, use the sample format shown below as a guide for converting your intangible needs into tangible, measurable goals. Figure 8.5 gives you a place to perform your own conversion of needs into goals. What you believe must affect your behavior, or it is doubtful that you really believe it.

1. Intangible need: personal growth

2. Conditions that would need to be present to satisfy your intangible need:
   a. College degree
   b. Profitable business
   c. At least x number of sales (in product or services)
   d. Kids grown and happy
   e. Communicate with spouse
   f. Leader in church
   g. Leader in community
   h. Leader in business

3. Converting each condition to a tangible goal:
   a. B.S. degree by June 1989

1. List your major intangible need or needs:
   a. _____
   b. _____
   c. _____
2. List the conditions that must be met in order for you to satisfy your intangible needs:
   a. _____
   b. _____
   c. _____
   d. _____
   e. _____
   f. _____
   g. _____
   h. _____
3. Convert each condition in item 2 to a tangible goal:
   a. _____
   b. _____
   c. _____
   d. _____
   e. _____
   f. _____
   g. _____
   h. _____

Figure 8.5 Converting needs to goals.

b. Twenty percent increase in business income in 1988

At least $x + y$ number of sales (in product or services)

d. College degrees for kids; $15,000 available for each child's education

e. One weekend a month alone with spouse

f. Sunday School teacher of young people's class

g. Officer in Jaycees

h. President of XYX National Businessmen's Association (or another organization representing your career interest)

Summary

The importance of goal setting in the achievement of life's goals cannot be overemphasized. The process of assessing personal and business goals, as described in this chapter, can help you to sort out conflicting priorities, to achieve greater self-understanding, to focus on beneficial actions and attitudes, and ultimately to find the satisfaction and success you crave.

9

The Electrical Contractor's Management Package

Accounting

Accounting and bookkeeping is by no means my field, but my experience tells me that you'll need the following:

1. Accounts receivable sales journal
2. Accounts receivable sales journal aging report
3. Month-to-date and year-to-date profit and loss (P and L) statements
4. Cash receipts report
5. Cash disbursements report
6. Accounts payable report
7. Trial balance sheet; monthly trial balance report
8. Balance sheet
9. Weekly job cost

In addition, you must make monthly adjustments to your over- and underbilling work-in-progress reports. These reports affect your profits.

Sample Balance Sheet

The balance sheet reveals the financial condition of a company as of a particular date. Operational results are reflected in this report, and a study of it can point up strong or weak areas in the financial structure. There are several ratios related to the balance sheet that can be used as guidelines. Examples of some of these ratios are listed below with explanations. Each ratio has a definite meaning; however, any one ratio, taken alone, may be inconclusive. Each ratio should be related to others to determine its significance. (See Table 9.1.)

Current Ratio. This ratio compares current assets (CA) with current liabilities (CL). It indicates a company's ability to pay its current debts from current assets. It also measures working capital available for day-to-day operations. A satisfactory ratio is considered to be 2:1, with a minimum of 1.8:1. Compute the current ratio by dividing current assets by current liabilities.

Your ratio[1] as of February 3, 1987, was 1.9:1. This means that you have $1.90 in assets for every $1 in liabilities, on a current basis.

Your position here is good, if your records correctly reflect it. The past-period ratio was 1.7:1. Your position then was OK, and it is important that you remain strong.

Acid Test, or Quick Ratio. This ratio is computed the same way as the current ratio, except that only assets that can be converted to cash within 30 to 60 days are included.

Your ratio here as of February 3, 1987, was 1.10:1. This would be considered satisfactory too; however, your position at the end of the previous period was 1.39:1. This area needs constant attention.

Ratio of Current Debt to Net Worth. This ratio compares the funds temporary creditors have invested with funds invested by owners. Net worth (NW)—defined as total assets less total liabilities—should be considerably greater to guarantee the creditor's investment. Compute the ratio of current debt to net worth by dividing current liabilities by net worth.

A ratio of less than 60 percent is considered safe. Your ratio was 1.02 percent as of February 3, 1987, and was 0.15 percent as of February 3, 1986. The 1987 ratio is above recommended limits and should be brought more in line.

Ratio of Total Debt to Net Worth. When total debt (current liabilities plus long-term debt) is related to net worth, the same reasoning as described above applies, except that this ratio includes funds *all* creditors have invested. Compute by dividing total debt by net worth. A ratio of less than 100 percent of net worth is considered safe.

Your ratio as of February 3, 1987, was 1.71 percent. Your position here needs improvement. This area should be *controlled.*

---

[1]All figures and dates in this chapter are cited for purposes of illustration only, and will not apply to readers' own companies.

Ratio of Net Fixed Assets to Net Worth. This ratio compares net dollars invested in fixed assets with net worth (owner's equity). It indicates the strength of net worth in terms of properties held for long-term use in producing income. Compute the ratio by dividing net fixed assets (NFA) by NW. A ratio of 30 to 50 percent is considered satisfactory.

This particular ratio reflects the character or makeup of your assets when related to net worth. It is considered prudent to have a low percentage of assets in *fixed* or low-producing categories. A low ratio reduces overhead (OH). It also keeps assets in comparatively liquid form.

Ratio of Net Sales to Net Worth. This ratio measures the turnover of owner's equity or net worth. A turnover of 6 to 9 times a year is considered satisfactory in most operations in this business because of low equity.

This ratio varies considerably depending upon the personal business habits of the contractor. Under some circumstances, a turnover of more than 8 times a year might be considered overextension for one contractor, whereas under other circumstances, the same turnover might be considered normal for another contractor. Net worth at the beginning of the period should be used. Unreasonable overextension— whatever "unreasonable" is—is always hazardous. In some operations, reasonable and controlled overextension might be considered normal for a particular company. Some people do not agree with this; however, the danger of overextension lies in the failure to put profit and an adequate amount of overhead on a job *into the estimate*. Thus, if a job goes sour, there is no provision for loss absorption, and the overextension can be a definite path to serious financial problems or insolvency.

Cashflow is a prime consideration here, and it is overlooked far too many times. Proper and adequate supervision is always important, but when overextension is present, *it is imperative*.

Ratio of Net Profit to Net Worth. This important ratio shows your return on investment: equity or net worth. This is not an operating ratio. Compute it by dividing net profit by net worth *at the beginning of period*. This ratio is expressed as a percentage, and acceptable percentages vary from 20 to 40 percent in this industry. It measures the profit you realized on your investment.

Ratio of Net Profit to Net Sales. This percentage measures your operating profit as a percent of net sales. The national average before taxes is about 4 percent. This average varies.

TABLE 9.1   Sample Electric Company Balance Sheet

Assets	1987	1986
Current assets		
Cash (on hand and in banks)	$ 24,868	$ 52,060
Accounts receivable	101,167	189,091
Loans to stockholder	80,366	617
Costs in excess of billing	13,901	0
Unbilled work	0	38,405
Materials inventory, at lower of cost or market	12,000	12,000
Prepaid expenses	5,316	5,316
Plan deposits	0	154
Total current assets	$237,618	$297,643
Fixed assets (at cost)		
Land	$ 55,000	$ 55,000
Building	35,000	35,000
Equipment and tools	16,899	10,659
Office equipment and furniture	14,265	14,265
Vehicles	26,858	26,858
	$148,022	$141,782
Accumulated depreciation	$ (52,704)	$ (42,676)
Total fixed assets (net)	$ 95,319	$ 99,106
Other assets		
Cash surrender value of insurance	$ 3,358	$ 2,073
Total other assets	$ 3,358	$ 2,073
Total assets	$336,294	$398,821
Liabilities and equity		
Current liabilities		
Current portion long term-debt	$ 21,283	$ 21,479
Notes payable to bank	25,000	6,500
Accounts payable	23,130	93,371
Other liabilities	300	0
Billings in excess of earnings	55,972	44,950
Income taxes payable	660	5,500
Total current liabilities	$126,345	$173,800
Long-term debt	$ 86,139	$111,495
Total long-term debt	$ 86,139	$111,495
Stockholder's equity		
Common stock, $10 par value, 10,000 shares authorized, 7977 shares issued and outstanding	$ 79,770	$ 79,770
Retained earnings (deficit)	33,757	(34,597)
Net income (loss)	10,283	68,354
Total equity	$123,810	$113,527
Total liabilities and equity	$336,294	$398,821

NOTE: This is a totally separate balance sheet for the early years of Sample Electric Company. It does *not* interface with the other tables in this chapter.

Accounts Receivable.   Receivables normally should not exceed about 6 to 8 percent of sales as of a particular date. Receivables should be aged monthly, to maintain a tight control—especially during tight times in the economy. (Aging is the multiple lengths of days that the accounts receivable and/or accounts payable are outstanding.) Aging should be done about the 7th or 12th of the month to reflect end-of-month billings and collections to present a realistic view of your situation.

Overhead

"Overhead" has been defined in many ways, and unfortunately, it means different things to different people. Overhead is that part of your business expense that continues whether you are currently working on any jobs or not. It is a continuous expense. It can be budgeted and controlled; therefore, not only can the dollar amount of overhead be estimated, but also, when estimating, recovery of overhead can be calculated for a given period or for a particular job. See Table 9.2 for overhead recovery by the interpolation method.

The significance of overhead is that it is always present and that it costs the contractor the same kind of dollars as those spent for productive labor, material, etc. It cannot be charged to direct job expense, and it must be recovered on each contract, whether as a lump-sum bid or on a cost-plus basis. *Overhead expense that is not recovered can only be absorbed from net profit.* (See Tables 9.3 and 9.4.)

All salaries or draws must be included in overhead burden. Any expense that cannot be charged directly to a job represents an overhead expense.

If you fail to include an adequate allowance for overhead in the estimate, or if, during the progress of a job, you incur actual dollar overhead expense, then this expense becomes unscheduled or unbudgeted overhead. It is absorbed from net profit on that job.

According to the Table 9.3 forecast, your overhead gross burden for the current fiscal period will be approximately $260,000. This means that, in addition to all direct job costs, you must recover this amount of overhead before you are in a profit area. To stress its importance to you, based on 252 working days per year, each working day you spend approximately $1,032 for overhead—and this happens whether you have any jobs or not. This amount has to be recovered or your net profit will be reduced to cover the deficiency.

You should keep in mind that overhead as a *percentage* will vary as the size of the contract or volume varies. The *larger* the contract, the *smaller* the *percentage* of overhead. It follows that the *smaller* the contract, the *larger* the *percentage* of overhead. This is important.

When overhead is excessive, you have three basic alternatives for adjustment: (1) increase sales without increasing overhead, (2) reduce overhead itself, or (3) increase the overhead provision for markup recovery.

It is important to know that overhead as a percentage of sales varies

TABLE 9.2   Overhead Adder:*†

No. of jobs	Size of job, $	Gross billing	Prime costs	% adder	$ adder
25	0–25	$ 325	$ 243	150	$ 364
30	25–50	1,050	787	100	787
20	50–100	1,500	1,125	100	1,125
15	100–200	2,250	1,687	95	1,109
10	200–300	2,500	1,875	98	1,837
20	300–500	7,500	5,625	97	5,456
10	500–1,000	7,500	5,625	96	5,400
8	1,000–2,000	12,000	9,000	95	8,550
7	2,000–3,000	17,500	13,125	48	6,300
5	3,000–5,000	20,000	15,000	45	6,750
4	5,000–10,000	28,000	21,000	30	6,300
1	10,000–25,000	12,500	9,375	6	562
1	25,000 and up	25,000	18,750	5	937
TOTAL		$137,625	$103,217	%	$45,477

No. of jobs	Size of job	Gross billing	Prime costs	% adder	$ adder
TOTAL					

*Dividing prime cost by 0.70 will cover overhead and leave $9827 to put back in business.
†This method of overhead recovery is primarily used when overhead is unknown, and it should be used by experienced accounting personnel only. Keeping track of the dollar size and number of jobs is of critical importance, for this table cannot be used without this information. The explanation for the first-line item, which also applies to all others, is as follows: 25 jobs times the midpoint of the $0–25 column ($12.50, or go to the next round dollar, which is $13.00) equals $325.00 gross billing. Prime cost is interpolated to be 75% of gross billing. The dollar adder is equal to 150% of prime cost.

inversely with sales. When sales are below the expected volume, overhead becomes an immediate problem. Overhead is not difficult to budget or control; it just takes constant planning and checking. We relate overhead as a percentage, but the only real consideration is how many dollars were spent and whether these dollars were recovered.

TABLE 9.3   Annual Sales, Costs, Overhead Requirements, and General Requirements for Sample Electric Company

Job	Sales required 2,200,000 (1)	Required OH $260,000 (12%) less recovery (2)	Contract days (3)	Daily OH recovery (4)	Material cost (5)	Labor cost (6)	DJE (7)	Truck available Y N (8)	L/D (9)	Work force E (10)	A	L	AHR (11)	Start Date (12)	Completion Date (13)
A	$ 795,820	$273,000	252	$1083	$287,660	$144,945	$ 83,795	1	55	3	1	2	13.98	10/ 1/85	7/23/86
B	179,739	20,000	120	90	30,591	26,200	88,700	1	35	2	1	1	14.00	10/ 1/85	1/15/86
C	93,159	16,168	180	74	8,658	21,436	46,895	1	100	1	1	1	11.00	10/15/85	2/28/86
D	98,923	13,275	180	213	44,584	31,376	9,687	1	50	1		3	12.00	10/30/85	3/ 1/86
E	190,605	54,844	180	113	75,713	0	59,450	1	0	1	1	2	15.00	10/15/85	4/30/86
F	26,000	1,695	15	60	16,379	1,533	1,772	1	2	1		2	11.07	11/15/85	11/30/86
G	27,663	5,407	90	23	15,875	5,872	506	1	10	1		1	13.00	10/16/84	11/25/86
H	15,995	2,788	120	88	8,025	5,000	100	(Worked simultaneously with job E)		1		1	15.00	10/15/85	2/15/86
I	100,365	15,896	180	111	37,452	43,417	3,600	1	25	2	1		14.00	10/30/85	4/30/86
J	318,457	4,220	30		319	4,055	26,686	1	25	1		1	13.00	10/30/85	11/30/86
Total sales to date	1,846,726	407,293	1347	1855	525,256	283,834	321,191	9	302	14	5	14	13.21		4.75 mo

NOTE: OH = overhead, DJE = direct job expense, LD = liquidated damages, AHR = average hourly rate. E= electrician. A = apprentice. L = laborer.

Also remember that, when overhead is measured as a percentage and when it is considered as a part of budgeting, "sales" means "billable sales."

1. The most urgent and pressing needs for any company are to exceed its breakeven point and to increase its sales. Such items as telephone bill, office expenses, and miscellaneous dues and subscriptions must be held in line and under control.

2. If you begin doing government contracts, you will need a person knowledgeable about government paperwork, and your field employees must be oriented to government work procedures before they go to the work site. These needs are critical to your profit.

3. A financial statement is needed every 3 to 6 months, and a profit and loss statement is needed *monthly*. These statements need not necessarily be done by a CPA, but perhaps in-house accounting is desirable. The reason for a monthly profit and loss statement is simple: You can make a 1-month correction and recover from it far sooner than you can from a 6-month error. (See Table 9.4. For purposes of illustration, the table can be used either as an example of a profit and loss statement or as a budget.)

Track labor (the most important job aspect) using Table 9.3:

Job	Months
A	9
B	3
C	4
D	5
E	6
F	6
G	(N/A)
H	4
I	1
Total	38 ÷ 8 = 4.75, or 4 3/4 months

Thirty-three production workers at 8 hours per day = 264 hours × average hourly rate $13.21 = $3,487.44 labor cost per day × 21 workdays per month = $73,236.24 monthly labor cost − $15,379.61 (21% labor tax) which is in direct job expense. This represents adjusted montly labor cost of $57,856.63 × 4.75 months = $274,818.99 total labor for all jobs.

Now look at column 6; for total labor $283,834, complete the work in 4 3/4 months and get a labor dollar surplus of $9,015.01. Based on $57,856.63 monthly labor divided by 21 workdays, you save $2,755.08 on labor cost for every day you finish ahead of schedule—and you earn $1,031.75 toward overhead, for a combined total of $3,786.83. The proper management of money, labor, and materials and tools can make this obtainable.

Column 8 shows availability of trucks needed.

Column 7 is direct job expense, measuring all items of cost related to the job including labor tax. "Labor tax" means tax on field labor. Administrative labor tax goes in overhead, because overhead is an item of expense that occurs whether you have a job or not.

Column 6 (labor cost), column 5 (materials cost), and column 7 (direct job expense) totals = prime cost of $1,130,281, which is 61 percent of total sales in column 1.

As for the overhead requirement of $260,000, at the top of column 2, $260,000 is 14 percent of $1,846,726, the total of sales to date in column 1:

	Percent*
Prime cost	61
Overhead	14
	75
DJE continuancy	17
Profit	+8
	100

*These figures represent a percentage of total sales to date in column 1.

Let me explain column 4 (daily overhead recovery) as it appears on the sales, cost, and general job requirements. There can be no profit until overhead is recovered.

OH recovered	$407,293
OH required	−260,000
OH surplus	$147,293

These were super jobs; as you can see, in 5 months you have recovered your overhead and and made an 8 percent profit. Sometimes this is impossible in an entire year. Column 4 clearly shows an overhead recovery on a daily basis and lets you know far enough ahead to prevent lack of overhead recovery.

1. Budget. (See Table 9.4.)

2. Annual sales, costs, and general overview. (See Table 9.3.)

3. Estimating.

The financial role of management—accounting. See Tables 9.1 to 9.5.

Budget and sales, cost, and overhead recovery analysis. The sales, costs, and general-requirements sheet (Table 9.3) cannot be put into perspective until your budget (Table 9.4, an early start-up budget for Sample Electric; it *does* interface with Table 9.3) has been made final.

Your budget tells you what areas of cost you project and how much volume you must do to (1) at worst, break even, or (2) make a profit. Column 2 shows a $260,000 overhead requirement divided by 252 working days; this equals $1,031.75. This is the amount you must spend every day just to operate administratively. Column 4 shows an overhead daily recovery figure of $1,855. You now show an $823.25 daily surplus for overhead. If this rate were to continue, you can well see what would happen: $823.25 × 252 = $207,459. Every electrical contractor would like to experience this; however, this is not practical because, according to the completion schedule, all jobs start and almost end at the same time some 4.75 or 5 months later. Also, according to Murphy's law, (1) nothing is as easy as it seems, (2) everything takes longer than you think it will, (3) if anything can go wrong, it will, and (4) it will go wrong at the worst possible time. Some jobs finish ahead of time, and others end up late and cost you liquidated damages, as in column 9.

According to column 10 (work force), the total 33 divided into the column 2 total $1,846,726 = $55,961. This represents the amount that each field employee must produce for billing over the given time period; or, simply stated, column 10 (electricians), represents crews (14) divided into column 1 equals $131,909 over the same given time frame, that every crew must produce.

Finally, column 1 indicates a sales goal of $2,200,000, of which $1,846,726 has been procured, with $353,274 in sales yet to procure. With 7 1/4 months yet to go, you can possibly achieve the sales, but the delay between job start-ups will not likely allow you to bill out the $353,274 before year end. This is rare in industry but is a possibility.

Example of Major Cost Breakdown

The P and L budget process needs constant attention because the constant changes that take place in the business operation affect the figures. An overbudget figure lasting for more than 4 months needs readjusting. Below is an elementary bid summary sheet (condensed). For a job this small, the markup would normally be higher.

Labor	$10,000
+ labor tax, 21%	2,100
+ material + tax	5,500
+ rental or other	500
+ direct job expense	1,000
= total prime cost	19,100
+ overhead, 10% of prime cost	1,910
= subtotal	21,010
+ profit + 10%	2,101
= subtotal	23,111
+ bond, 1.5% of 23,111	+ 347
= selling price	23,458

Note Overhead should be figured as a percentage of prime cost.

The budget must be monitored monthly. The secret of success is to get the sales volume. Your ratios indicate why the banks will or will not work with you in a positive manner.

According to Table 9.3, your prime cost must be divided by 0.51 to obtain the sales figure.

Based on Table 9.3, your total cost is $1,390,281 including overhead. This means you must bill and collect $115,856 monthly just to break even. In other words, this is your breakeven point on prime cost and overhead. Should you not have any jobs, then you won't have prime cost. Therefore, deducting $1,130,281, from total cost leaves you $260,000 worth of overhead. Your breakeven point on overhead only is $260,000; divided by 12, this equals $21,666, the monthly billing you must maintain in order to stay in business and pursue sales. Again, I remind you that in order to keep track of the financial management of your business, you and your CPA or accountant must follow and control the budget.

Prime Cost: Definition. The three items that make up *prime cost* are as follows:

1. *Direct job expense.* Any expense related to a job, that is, any expense you would not have if you were not doing that job. Examples are subcontracts, rentals, small-tool purchases, field payroll and fringe benefits paid to field employees, payroll taxes, permits, labor taxes, public liability and property damages, and job-related insurance.

2. *Material costs.* All materials purchased for a specific job or installation.

3. *Labor costs.* All labor costs incurred during installation and handling of materials for a job.

In other words,

Direct job cost + labor + material + tax = prime cost

In the electrical industry it is standard practice to mark up your prime cost to cover overhead, but you cannot do this until you know what your overhead (shown in the sales forecast in Table 9.3) is.

Potential Areas of Damages

Set out below is an outline of potential areas of damages in delay claims. This outline can be used as a guide in calculating delay damages.

I. *Time-related damages resulting from noncontractor-caused delays.*
   A. *Extended equipment rentals or depreciation.*
   B. *Extended job site overhead.* This includes all items directly allocated to the job, such as field engineering, clerical personnel,

storage and maintenance, insurance at site, site security, site utilities, and trailers.

   C. *Extended home-office overhead.* Base the computation on the Eichleay formula (which is customarily used by the U.S. government and by contractors to determine a fixed cost for delays not caused by the contractor):

     1. *Contract (job) billing.*
Total billings for contract period × total overhead for contract period = overhead allocable to contract (job)

     2. *Allocable overhead.*
Days for performance = daily contract overhead

     3. *Daily contract.*
Overhead × number of days of delay = amount to claim for home office overhead

  II. *Cost inflation or escalation.*

   A. Labor rates bid versus labor rates paid in delay period.

   B. Material costs bid versus material costs in delay period.

 III. *Loss of labor efficiency.*

   A. *Labor inefficiency.* Inefficient work performance may be caused by:

     1. Less than efficient crews resulting from non-contractor-caused delay

     2. Work performed out of sequence

     3. Work shifted from good weather to bad weather

     4. Cost of shutdown and startup

     5. Congestion in work areas

     6. Rehandling of material

   B. *Base inefficiency factor.* This factor (which is the base rate figured for production before the delay) may come into play on labor costs per unit from other jobs in the same area versus labor cost per unit for this job. If such figures are not available, utilize standard industry indexes.

 IV. *Specific claims for extra labor and material.*

   A. *Cost of work or material.*

   B. *Profit and overhead.*

  V. *Loss of anticipated profit from other jobs.* Amount of profit which could have been made on other projects, had the delay not occurred.

 VI. *Interest on claims.* The Eichleay formula can be used in filing a legal claim for a job delay not caused by the contractor. The contractor seeks to charge interest on the claim amount until is is paid, once an agreement has been reached.

 VII. *Field directives and extra work orders which have not been settled.*

VIII. *Owner-furnished materials.* Figure 9.1 (see chapter appendix) gives an example of a letter advising clients of possible extra costs associated with use of owner-furnished materials.

Cashflow

When dealing with huge contracts or sales volume, or both, one of the most important considerations is cashflow. It is important that your line of credit be established in an amount that is sufficient and that it be established before the need may arise. When cashflow becomes strained, you become more and more reliant on the bank and they know it.

Should you appear to the bank to be a possible liability, they might start pressuring you and begin to dictate operational procedures to you. It is important to set a reasonable, profitable, and controllable limit to your growth so that this situation will not develop beyond your control. Cashflow can be affected in several ways (see Tables 9.5 and 9.6):

1. Loss of net profit.

2. Failure to control spending of cash.

3. Overextension, which is dangerous even if profit is good. Cash can flow into nonliquid areas, and something could happen that might not be your fault but would jam your cashflow.

4. Excess tie-up in inventory.

5. Excess work in progress.

6. Excess cash invested in alien assets.

7. Ignoring financial and operating ratios.

8. Too much cash tied in fixed assets or low-producing categories.

9. Underbilling.

10. Overbilling, uncontrolled operations, and early net profit; then diversion of cash into other categories.

11. Heavy billing for material (during uncertain times in the economy) that is not installed until later, and then leaking cash into the wrong areas.

12. Allowing too much retainage or trade receivables, or both, to accumulate, especially when overextension is present. *This is very important when there is a lull in the economy.*

13. Not knowing or not checking on general contractors at bid level. Two general contractors may not get the same price from you. Try

TABLE 9.4   Sample Electric Company Budget

	Annual budget	%	Monthly budget
Net sales	$160,000	100	$13,333
Prime cost			
Material	62,400	39	5,200
Labor*	44,800	28	3,733
Direct job expense†	8,000	5	666
Total prime cost	115,200	72	9,600
Gross profit	44,800	28	3,733
Overhead expense‡			
Advertising	200		17
Automobile and truck	1,500		125
Bad debts	600		50
Charity donations	200		17
Cost of collections	300		25
Depreciation	300		25
Dues and subscriptions	350		29
Employee benefits	400		33
Freight and postage	80		7
Heat, light, and power	350		29
Interest	1,000		83
Insurance	1,500		125
Workers' compensation, public liability, and property damage insurance included in DJE§		–0–	
Legal and accounting	800		67
Misc. expenses	500		42
Office supplies and expenses	300		25
Pension and welfare contributions	200		17
Callbacks	500		42
Rent	3,500		292
Repair and maintenance	200		17
Salaries	15,000		1,250
General administrative			
Engineering and estimating	–0–		–0–
Nonproductive labor	–0–		–0–
Office and clerical	3,000		250
Truck and warehouse	–0–		–0–
Sales and commission selling	–0–		–0–
Selling expenses	–0–		–0–
Shop supplies and expenses	–0–		–0–
Small tools	–0–		–0–
Taxes and license	600		50
Payroll taxes (office only)	1,000		83
Telephone and telegraph	600		50
Travel and entertainment	–0–		–0–
Consulting fees, seminars	3,000		250
Total overhead expense	35,980	22	3,000
Operating net profit overhead actually recovered	8,820	6	733

*Based on 252 workdays per year = $143.00 overhead per day.
†$143.00 per day or $17.30 per hour.
‡You must do $149,980 just to break even, or $12,498 per month sales.
§Based on $115,200 of prime cost; you must divide it by 0.70 = $164,571 worth of selling to cover all expenses and make a profit.

TABLE 9.5 Schedule of Position on Completed and Uncompleted Contracts*

Job	(1) Original contract price	(2) Original estimated cost	(3) Revised contract price	(4) Revised estimated cost	(5) Total cost to date	(6) Estimated cost to complete	(7) % complete	(8) Billed to date	(9) Revenue earned to date	(10) Billings over (under)	(11) Estimated gross profit	(12) % gross profit
A	272,157.00	192,253.00	222,157.00	214,816.00	210,216.00	4,500.00	98	210,667.00	217,713	(7,046)	7341.00	3.30
B	456,034.00	358,711.00	506,049.00	422,390.00	417,772.00	4,618.00	99	474,396.00	500,988	(26,592)	83,659.00	16
C	324,888.00	302,618.00	362,112.00	380,000.00	374,367.00	5,633.00	99	339,541.00	358,491	18,950	17,888.00	Loss
D	47,300.00	42,253.00	47,300.00	46,800.00	45,876.00	924.00	98	43,201.00	46,354	(3,153)	500.00	1
E	49,000.00	48,680.00	59,000.00	55,000.00	40,632.00	14,368.00	74	29,124.00	43,660	(14,536)	4,000.00	7
F	795,820.00	503,528.00	795,820.00	503,530.00	13,540.00	489,990.00	3	13,540.00	23,875	(10,335)	292,290.00	36
G	179,738.00	139,189.00	179,738.00	139,189.00	-0-	139,189.00	-0-	-0-	-0-	-0-	40,549.00	5
H	93,159.00	72,584.00	93,159.00	72,584.00	-0-	72,584.00	-0-	-0-	-0-	-0-	20,575.00	2.20
I	98,923.00	58,282.00	98,923.00	58,282.00	-0-	58,282.00	-0-	-0-	-0-	-0-	40,641.00	41
J	190,605.00	130,946.00	190,605.00	130,946.00	25,244.00	105,707.00	19	-0-	36,215	(36,215)	59,659.00	31.2
K	25,660.00	20,239.00	25,660.00	20,239.00	-0-	20,239.00	-0-	-0-	-0-	-0-	5,421.00	21
L	27,663.00	25,472.00	27,663.00	25,472.00	25,472.00	-0-	100	27,663.00	27,663	-0-	2,191.00	8
M	15,995.00	11,453.00	15,995.00	11,453.00	3,320.00	8,133.00	29	-0-	4,638	(4,638)	4,542.00	28
N	100,342.00	77,128.00	100,342.00	73,128.00	34,723.00	38,405.00	47	42,388.00	47,160	(4,772)	27,214.00	27
O	31,475.00	27,770.00	31,475.00	27,770.00	-0-	27,770.00	-0-	-0-	-0-	-0-	3,705.00	12
P	138,475.00	102,117.00	138,475.00	102,117.00	-0-	102,117.00	-0-	-0-	-0-	-0-	36,358.00	2
	2,847,234.00	2,113,223.00	2,894,473.00	2,283,716.00	1,191,162.00	1,092,554.00		1,180,520.00	1,306,757	(126,237)	610,757.00	

*Column 5 – column 4 = column 6; but you must take the contractor's word, and the figure may be plus or minus. Column 5 ÷ column 4 = column 7. Column 8 comes directly off the contractor's billing schedule. Column 7 × column 3 = column 9. Column 9 – column 8 = column 10. Column 3 – column 4 = column 11.

TABLE 9.6 Cashflow Sheet

	Total	Oct.	Nov.	Dec.	Jan.	Feb.	Mar.	Apr.	May	June	July	Aug.	Sept.
1 Contract days	300												
2 Actual workdays	216	23	19	23	20	21	23	22	21	23	21	21	22
3 Labor load 6; hours	10,368	1104	912	1104	960	1008	1104	1056	1008	1104	1008		
4 Contract amount	$795,820												
5 Material	$287,660		$28,000*		$255,360†	$1,000	1,000	500	800	1,000	$1,000		
6 Labor	144,945	15,434	14,092	12,750	15,434	13,420	14,092	15,434	14,763	14,092	15,434		
7 DJE	84,795	28,800	6,159	5,877	6,441	6,018	6,159	6,441	6,300	6,159	6,441		
8 Discount			560		5,087	20	20	10	16	20	20		
9 Total cash used	517,400	44,234	48,251	18,627	276,235	20,438	21,251	22,375	21,863	21,251	22,875		
10 Billing	795,820	72,076	76,093	409,588	40,000	33,000	33,013	33,013	33,013	33,013	33,011		
11 Receipts		0	0	133,352	368,629	36,000	29,700	29,712	33,013	33,013	33,013	33,012	66,376
12 Cash (need) surplus = line 11 − line 9		(44,234)	(48,251)	114,725	92,394	15,562	8,449	7,337	11,150	11,762	10,138		
13 Profit = line 4 − line 9 = line 14													
14	278,420												278,420
15 Includes: bond, crane, labor tax, dump truck, barricade, duct, rodder (fuel maintenance).													
16 Curtail loan				60,000	30,000	2,485							

NOTE: When job is 50% complete, 1/2 of 10% retainage is released, or if job goes on or ahead of schedule, no retainage is held.
*Splice and terminations.
† Receive wire only.

to procure only jobs on which you yourself can be the general contractor, so that you will be able to control the job and the money.

14. Lack of cost controls.

15. Taking on contracts that involve retainage (a percentage of money that may be deducted from each bill the contractor submits) for too long a time, causing additional strain on capital.

16. Incorrect estimates.

17. Permitting credits due you from supply houses to lag too long. A good way to avoid this is to not permit suppliers to owe you credits—to take credits when they are due.

18. Having many company areas that contribute to poor production. This includes activities or administrative management.

19. Failure to treat cash as a commodity—to shop around for the best interest rate and the best banking deal available.

20. Failure to plan cashflow needs ahead of time.

21. Improper professional management of your business.

Job-Cost Control

There are two major specialities in accounting: *taxes* and *costing*. Both are important to you; taxes are necessary, and cost accounting will help control costs and profits. When your operating figures reach the profit and loss stage, it's too late for controls. See Table 9.7 for a commercial job-cost report and Table 9.8 for a residential job-cost report. The contractor should match actual costs against the estimate. Estimated versus actual costs are clearly shown in Table 9.6, though overhead is yet to be deducted from the profit-to-date figure.

It is important for you to have control of a job at every stage, starting with the estimate, during the process of the work, and at completion. See Table 9.6 for this control. For tool, equipment, and inventory control, see Table 9.8. For daily control, see Table 9.9. Tables 9.10 to 9.12 are samples for billing by unit installation placement and certified payrolls.

In order to maintain proper job-cost control, a *unit-cost analysis report* should be done regardless of the type of accounting system you use; even if you are mechanized or computerized, the information should be maintained on cost cards. *Materials costs should be posted daily if possible, and labor costs should be posted at least weekly.* Direct job costs should be recorded as they are charged to a job. Progress billings can be made from the job-cost analyst report if desired. (If

TABLE 9.7 Job-Cost Report*

Sample Electric Company

Job 101	Client	Type 6	Square Feet Material Estimator		Contract Ch Order Total	181,880.00 0.00 181,880.00	Billed Received Retained	177,308.00 170,746.55 0.00
05-21-87	Completion Date							

Run On 02-28-87   0.00

Phase	Estimated			Current			To Date			% comp.	Hours re-maining	$ to comp.
	Hours	Cost/hour	Cost	Hours	Cost/hour	Cost	Hours	Cost/hour	Cost			
03 BR			E									
Labor	1,010.50	$9.43	9,529.01	—	—	—	1,101.50	$ 9.03	$ 9,950.76	104	91	$ 421.75
Material	—	—	5,960.44	—	—	$ 21.46	—	—	5,960.44	100	0	0.00
PHASE TOTAL	1,010.50	9.43	15,489.45	—	—	21.46	1,101.50	9.03	15,911.20	103	91	421.75
04 BW												
Labor	320.00	9.43	3,017.60	—	—	—	320.00	12.53	4,011.05	133	0	993.45
Material	—	—	5,631.39	—	—	366.93	—	—	5,631.39	100	0	0
PHASE TOTAL	320.00	9.43	8,648.99	—	—	366.93	320.00	—	9,624.44	112	0	993.45
05 C												
Labor	64.00	9.43	603.52	—	—	—	54.00	12.69	685.47	114	10	81.95
Material	—	—	743.13	—	—	156.78	—	—	743.13	100	0	0
PHASE TOTAL	64.00	9.43	1,346.65	—	—	156.78	54.00	12.69	1,428.60	106	10	81.95
06 Cement												
Material	—	—	14,307.18	—	—	68.52	—	—	1,407.18	100	0	0.00
PHASE TOTAL	—	—	14,307.18	—	—	68.52	—	—	1,407.18	100	0	0.00
08 F												
Labor	440.00	9.43	4,149.20	—	—	—	439.00	12.87	5649.42	136	1	1,500.22
Material	—	—	38,025.00	—	—	—	—	—	36,534.86	96	0	1,490.14
PHASE TOTAL	440.00	9.43	42,174.20	—	—	—	439.00	12.87	42,184.28	100	1	10.00
10 Cutting and removal of road												
Labor	8.00	9.43	75.44	—	—	—	8.00	4.36	34.88	46	0	40.56
PHASE TOTAL	8.00	9.43	75.44	—	—	—	8.00	4.36	34.88	46	0	40.56

	Hours	Rate	Amount	Hours	Rate	Amount	Hours	Rate	Amount			Amount
Phase	**11 Seeding landscape**											
Labor	252.00	9.43	$2,376.36	100.00	$13.20	$1,319.84	252.00	$11.89	$2,997.04	126	0	$620.68
PHASE TOTAL	252.00	9.43	2,376.36	100.00	13.20	1,319.84	252.00	11.89	2,997.04	126	0	620.68
Phase	**12 Test and inspect packing list**											
Labor	64.00	9.43	603.52	8.00	12.33	98.64	8.00	12.33	98.64	16	56	504.88
PHASE TOTAL	64.00	9.43	603.52	8.00	12.33	98.64	8.00	12.33	98.64	16	56	504.88
Phase	**13 Demolition**											
Labor	110.00	9.43	1037.30	112.00	13.59	1,522.56	112.00	13.59	1,522.56	147	2	485.26
PHASE TOTAL	110.00	9.43	1037.30	112.00	13.59	1,522.56	112.00	13.59	1,522.56	147	2	485.26
Phase	**14 Demobilization**											
Labor	32.00	9.43	301.76	22.00	12.33	271.26	22.00	12.33	271.26	90	10	30.50
PHASE TOTAL	32.00	9.43	301.76	22.00	12.33	271.26	22.00	12.33	271.26	90	10	30.50
Phase	**15 Misc.**											
Labor	29.00	9.43	273.47	29.00	12.33	357.57	29.00	12.33	357.57	131	0	84.1
Material	—	—	1,889.28	—	—	21.46	—	—	1,889.28	100	0	0
PHASE TOTAL	29.00	9.43	2,162.75	29.00	12.33	357.57	29.00	12.33	2,246.85	104	0	0
Phase	**16 DJE bond**											
Other	—	—	3,092.40	—	—	—	—	—	3,092.40	100	0	0
PHASE TOTAL	—	—	3,092.40	—	—	—	—	—	3,092.40	100	0	0
Phase	**17 DJE rentals**											
Other	—	—	1,573.10	—	—	324.22	—	—	1,824.78	116	0	251.68
PHASE TOTAL	—	—	1,573.10	—	—	324.22	—	—	1,824.78	116	0	251.68

TABLE 9.7 Job-Cost Report *(Continued)*

SAMPLE ELECTRIC COMPANY

Job 101	Client	Type 6	Square Feet	0.00	Contract	181,880.00	Billed	177,308.00
	Completion Date		Material		Ch Order	0.00	Received	170,746.55
05-21-87			Estimator		Total	181,880.00	Retained	0.00

RUN ON 02-28-87

	Estimated			Current			To Date			% comp.	Hours re-maining	$ to comp.
	Hours	Cost/hour	Cost	Hours	Cost/hour	Cost	Hours	Cost/hour	Cost			
Phase 18 DJE Misc.												
Other	—	—	$ 7,538.00	—	—	—	—	—	$ 2,547.16	34	0	$4,990.84
PHASE TOTAL	—	—	7,538.00	—	—	—	—	—	2,547.16	34	0	4,990.84
Phase 20 DJE boring, cut, patch												
Other	—	—	6,601.00	—	—	—	—	—	5,991.00	91	0	610.00
PHASE TOTAL	—	—	6,601.00	—	—	—	—	—	5,991.00	91	0	610.00
Phase 21 Mobilization												
Labor	86.00	$9.43	810.98	—	—	—	86.00	$ 8.83	759.39	94	0	51.59
PHASE TOTAL	86.00	9.43	810.98	—	—	—	86.00	8.83	759.39	94	0	51.59
Job total												
Labor	2,415.50	9.42	22,778.16	271.00	$13.17	$3,659.87	2,431.50	10.83	26,338.04	116	16	3,559.88
Material	—		53,656.42	—		613.69	—		52,166.28	97	0	1,490.14
Maintenance	—		—	—		—	—		—	0	0	0
Equipment	—		—	—		—	—		—	0	0	0
Subcontract	—		—	—		—	—		—	0	0	0
Other	—		18,804.50	—		324.22	—		13,455.34	72	0	5,349.16
	2,415.50	9.43	95,239.08	271.00	13.17	4,507.78	2,431.50	10.83	91,959.66	97	16	3,279.42

*The meanings of the following phrases are: BR = feeder conduit rough-in; BW = feeder wire; C = panels; D = fixtures, or poles fixtures included The remainder of the cost codes are self-explanatory.

TABLE 9.8 Labor and Cost Estimates for Various Types of Houses

Type of house	First phase, hours	Second phase, hours	Third phase, hours	Fourth phase, hours	Fifth phase, hours	Total hours
A. Ranch	3	12		12	15	42
B. Bilevel	2	9	11	20	16	58
C. Bilevel	3	9	8	14	12	46
D. Ranch 700	3	12		12	15	42
E. Bilevel 500	3	9	11	20	16	59
F. Colonial 800	3	8	16	17	14	58
G. Split-level 601	3	14	12	14	13	56
H. Ranch 509F	3	12		12	15	42
I. Ranch 505	3	8		8	12	31
J. Ranch 700A	3	12		12	15	42

Type of house	Labor cost (3.3)	Labor overhead (3.0)	Material cost	Other direct cost (percent)	Total cost	Profit	Selling price
A. Ranch	$273.20	$252.00	$642.62	$16.00	$1183.82	$104.78	$1288.60
B. Bilevel	389.40	354.00	770.00	21.00	1534.4	(79.52)	1454.88
C. Ranch, 4-bedroom	336.60	306.00	686.30	33.00	1361.90	95.30	1457.20
D. Bi-level	297.00	270.00	472.00	38.00	1077.00	470.60	1547.60
E. 2-story colonial	382.80	348.00	753.88	42.00	1526.68	336.92	1863.60
F. Split-level	389.60	336.00	680.00	38.00	1443.60	214.76	1658.36
G. Ranch 700	277.20	252.00	720.00	33.00	1282.00	357.80	1640.00
II. Bi-level 500	389.40	354.00	740.00	38.00	1521.40	258.80	1780.20
I. Colonial 800	382.80	348.00	754.00	42.00	1526.80	457.82	1984.62
J. Split level 601	369.60	336.00	680.00	38.00	1423.60	234.76	1658.36

progress billings are permitted only on a percentage-of-completion basis, then adjust your entries accordingly.)

Since direct labor is highly variable, you need *weekly information concerning labor productivity on each job.*

In addition to unit job control, management needs a composite report on a monthly basis which shows all costs for each job in progress. This can be accomplished by preparing a *monthly job-cost summary report.* This report can be prepared by the 5th of each month or sooner if desired.

To know the total costs of a particular job to date is not enough: The question is, are these costs *good* or *bad*? Costs are judged good or bad by comparing them with the estimate, as shown in Table 9.7. These comparisons should be by cost category (materials, labor, and direct job expense), not just by total. In many instances the categories should be broken down further to be more effective. By comparing direct job costs with the estimated time or costs, you can determine how much of the estimated time or costs have been used to date. Then, by further comparing these figures with the percentage of the job that is complete, you can develop some valuable cost information. The column "%

TABLE 9.9 Tool, Equipment, and Inventory Control

Inventory Stock Status Report (By Item Number)
Inventory Id: Stock   Title: Main A/R Inventory

	On hand	On Order	Mini-mum	Maxi-mum	Weight	Cost	Sell*	Profit*	MTD sales	YTD sales
Item 01 Desc.: Ralph Vend. Loc.: Newport News · Product code IDN · License No. XYC-364 · Unit of measure E 82 S-10 PU truck	1	0	0	0	0	6,800.00	0.00	Total cost — Total resale value $6,800.00 · 100.00%	6,800.00 · 0	0
Item 01HV Desc.: PLM penciling tool Vend. Loc.: Larry Trk Trl · Product code · PT 10B.295 · Unit of measure E	1	0	0	0	0	67.00	0.00	Total cost — Total resale value 67.00 · 100.00	67.00 · 0	0
Item 01T Desc.: 12-lb mall Vend. Loc.: Job 90 · Product code · Unit of measure	1	0	0	0	0	33.40	0.00	Total cost — Total resale value 33.40 · 100.00	33.40 · 0	0
Item 02 Desc.: George Vend. Loc.: Portsmouth · Product code IDN · License No. XXZ-747 · Unit of measure E	1	0	0	0	0	6,800.00	0.00	Total cost — Total resale value 6,800.00 · 100.00	6,800.00 · 0	0
Item 02HV Desc.: Screw-type schackles Vend. Loc.: Larry Trk Trl · Product code · Unit of measure E	5	0	0	0	0	16.80	0.00	Total cost — Total resale value 16.80 · 100.00	84.00 · 0	0
Item 02T Desc.: Pick Vend. Loc.: Job 90 · Product code · Unit of measure	1	0	0	0	0	20.18	0.00	Total cost — Total resale value 20.18 · 100.00	20.18 · 0	0

Item 03 Desc.: Larry Ross Product code License No. XYC-364 80 navy blue Chevy Van
Vend. Loc.: Daleville Unit of measure E Total cost 2,500.00 Total resale value 2,500.00

| 1 | 0 | 0 | 0 | 0 | 2,500.00 | 0.00 | 100.00% | 0 | 0 |

Item 03HV Desc.: 2-ton Lugail hot stick hoist 4 Product code
Vend. Loc.: Larry Trk Trl Unit of measure 1 Total cost 178.00 Total resale value 178.00 178.00

| 1 | 0 | 0 | 0 | 0 | 178.00 | 0.00 | 100.00 | 0 | 0 |

Item 03T Desc.: Digging bar Product code Unit of measure E
Vend. Loc.: Job 90 Total cost 34.94 Total resale value 34.94 34.94

| 1 | 0 | 0 | 0 | 0 | 34.94 | 0.00 | 100.00 | 0 | 0 |

Item 04 Desc.: Ron Product code License No. DZB-841 82 yellow S-10 Chevy PU
Vend. Loc.: Newport News Unit of measure E Total cost 5,500.00 Total resale value 5,500.00 5,500.00

| 1 | 0 | 0 | 0 | 0 | 5,500.00 | 0.00 | 100.00 | 0 | 0 |

Item 04HV Desc.: 2" X 6" Liftan Sling Bylon Product code Larry Trk Trl
Vend. Loc.: Unit of measure F Total cost 36.40 Total resale value 36.40 36.40

| 1 | 0 | 0 | 0 | 0 | 36.40 | 0.00 | 100.00 | 0 | 0 |

Item 04T Desc.: Round-point shovels Product code Unit of measure E
Vend. Loc.: Job 90 Total cost 34.89 Total resale value 34.89 69.78

| 2 | 0 | 0 | 0 | 0 | 34.89 | 0.00 | 100.00 | 0 | 0 |

TABLE 9.9 Tool, Equipment, and Inventory Control *(Continued)*

INVENTORY STOCK STATUS REPORT   (BY ITEM NUMBER)
INVENTORY ID: STOCK   TITLE: MAIN   A/R INVENTORY

On hand	On Order	Mini-mum	Maxi-mum	Weight	Cost	Sell*	Profit*		MTD sales	YTD sales
Item 05	Desc.:									
Vend. Loc.:				License No. DDL-806			84 Chevy S-10 PU			
		Product code		Unit of measure E			Total cost			
1	0	0	0	0	6,759.00	0.00	6,759.00		6,759.00	
							Total resale value			
							6,759.00	100.00%	0	0
Item 05HV	Desc.: 4800-watt generator									
Vend. Loc.: Larry Trk Trl				#178A50						
		Product code		Unit of measure E			Total cost		1,424.00	
1	0	0	0	0	1,424.00	0.00	1,424.00			
							Total resale value			
							1,424.00	100.00	0	0
Item 05T	Desc.: Square-point shovels									
Vend. Loc.: Job 90										
		Product code		Unit of measure E			Total cost		69.78	
2	0	0	0	0	34.89	0.00	34.89			
							Total resale value			
							34.89	100.00	0	0
Item 06	Desc.:									
Vend. Loc.:				License No. EFC-559			73 Chevy C-10 blue GMC			
		Product code		Unit of measure E			Total cost		800.00	
1	0	0	0	0	800.00	0.00	800.00			
							Total resale value			
							800.00	100.00	0	0
Item 06HV	Desc.: Pumps									
Vend. Loc.: Larry Trk Trl				#12053						
		Product code		Unit of measure E			Total cost		6,132.00	
2	0	0	0	0	3,066.00	0.00	3,066.00			
							Total resale value			
							3,066.00	100.00	0	0
Item 06T	Desc.: Push brooms									
Vend. Loc.: Job 90										
		Product code		Unit of measure E			Total cost		24.00	
2	0	0	0	0	12.00	0.00	12.00			
							Total resale value			
							12.00	100.00	0	0

Item 07 Desc.: C. L. Ray Jr., Inc License No. WWV-441 73 International Scout
Vend. Loc.: Norfolk Product code Unit of measure E Total cost 800.00 800.00
                                                      Total resale value 100.00% 0
                                                                                    0

1	0	0	0	800.00	0.00	800.00

Item 07HV Desc.: Hoses #465744
Vend. Loc.: Larry Trk Trl Product code Unit of measure E Total cost 369.00 738.00
                                                            Total resale value 100.00 0
                                                                                          0

2	0	0	0	369.00	0.00	369.00

Item 07T Desc.: Cement trowels
Vend. Loc.: Job 90 Product code Unit of measure E Total cost 12.00 24.00
                                                     Total resale value 100.00 0
                                                                                   0

2	0	0	0	12.00	0.00	12.00

Item 08 Desc.: Ed License No. ENH-614 78 Dodge red and white PU
Vend. Loc.: Norfolk Product code Unit of measure E Total cost 2,613.50 2,613.50
                                                       Total resale value 100.00 0
                                                                                     0

1	0	0	0	2,613.50	0.00	2,613.50

Item 08HV Desc.: Hoses #465744
Vend. Loc.: Larry Trk Trl Product code Unit of measure E Total cost 183.60 367.20
                                                            Total resale value 100.00 0
                                                                                          0

2	0	0	0	183.60	0.00	183.60

Item 08T Desc.: Pair 36-in. bolt cutters No. 3
Vend. Loc.: Job 90 Product code Unit of measure Total cost 60.00 60.00
                                                   Total resale value 100.00 0
                                                                                 0

1	0	0	0	60.00	0.00	60.00

TABLE 9.9 Tool, Equipment, and Inventory Control *(Continued)*

INVENTORY STOCK STATUS REPORT   (BY ITEM NUMBER)
INVENTORY ID: STOCK   TITLE: MAIN A/R INVENTORY

On hand	On Order	Mini-mum	Maxi-mum	Weight	Cost	Sell*	Profit*	MTD sales	YTD sales
Item 09 Desc.: C.L. Ray Jr., Inc.				License TG 62-072			70 Ford bucket truck		
Vend. Loc.: Virginia Beach			Product code	Unit of measure E			Total cost		
							Total resale value		
1	0	0	0	0	7,477.00	0.00	7,477.00 100.00%	7,477.00	0
								0	0
Item 09HV Desc.: Gastech protector, combustible									
Vend. Loc.: Larry Trk Trl			Product code	Unit of measure 1			Total cost		
							Total resale value	875.00	
1	0	0	0	0	875.00	0.00	875.00 100.00		0
								0	0
Item 09T Desc.: 5 cu. ft. wheelbarrow M-11 5									
Vend. Loc.: Job 90			Product code	Unit of measure			Total cost		
							Total resale value	77.00	
1	0	0	0	0	77.00	0.00	77.00 100.00		0
								0	0
Item 10 Desc.: C.L. Ray Jr., Inc.				License FWZ-698			84 blue Jeep Cherokee		
Vend. Loc.: Virginia Beach			Product code	Unit of measure E			Total cost		
							Total resale value	16,000.00	
1	0	0	0	0	16,000.00	0.00	16,000.00 100.00		0
								0	0
Item 10HV Desc.: Power dart				#80230					
Vend. Loc.: Larry Trk Trl			Product code	Unit of measure 1			Total cost		
							Total resale value	789.00	
1	0	0	0	0	789.00	0.00	789.00 100.00		0
								0	0
Item 10T Desc.: 50-ft. garden hoses									
Vend. Loc.: Job 90			Product code	Unit of measure E			Total cost		
							Total resale value	64.00	
8	0	0	0	0	8.00	0.00	8.00 100.00		0
								0	0

Item 11   Desc.:: Larry   Product code   License FMU-120   79 Chevy 1/2-ton PU, blue
Vend. Loc.: Chesapeake   Unit of measure E

								Total cost		3,640.00			
								Total resale value		3,640.00	100.00%	0	0
1	0	0	0	0	3,640.00	0.00	3,640.00						

Item 11HV   Desc.: Manhole hooks   Product code   #80230
Vend. Loc.: Larry Trk TRL   Unit of measure 1

								Total cost		22.56		45.12	
								Total resale value		22.56	100.00	0	0
2	0	0	0	0	22.56	0.00	22.56						

Item 11T   Desc.: Ridged strap wrenches   Product code   6" Capacity #C-36
Vend. Loc.: Job 90   Unit of measure E

								Total cost		31.90		63.00	
								Total resale value		31.90	100.00	0	0
2	0	0	0	0	31.90	0.00	31.90						

Item 12   Desc.:   Product code   License TG 94-863   68 Line truck
Vend. Loc.:   Unit of measure E

								Total cost		6,000.00		6,000.00	
								Total Resale value		6,000.00	100.00	0	0
1	0	0	0	0	6,000.00	0.00	6,000.00						

Item 12HV   Desc.: PLM penciling tool   Product code   P.T. 750D .295
Vend. Loc.: Larry Trl   Unit of measure E

								Total cost		67.00		67.00	
								Total Resale Value		67.00	100.00	0	0
1	0	0	0	0	67.00	0.00	67.00						

Item 12T   Desc.: Traffic cones   Product code   Unit of measure E
Vend. Loc.: Job 90

								Total cost		8.00		280.00	
								Total resale value		8.00	100.00	0	0
35	0	0	0	0	8.00	0.00	8.00						

TABLE 9.9 Tool, Equipment, and Inventory Control (Continued)

Inventory Stock Status Report   (By Item Number)
Inventory Id: Stock   Title: Main A/R Inventory   D

On hand	On Order	Mini-mum	Maxi-mum	Weight	Cost	Sell*	Profit*		MTD sales	YTD sales
Item 13 Desc.: C. L. Ray Jr., Inc.										
Vend. Loc.: Norfolk			Product code		License FFR-560 Unit of measure E		74 winch truck Total cost Total resale value			
1	0	0	0	0	8,500.00	0.00	8,500.00	100.00%	8,500.00	0 0
Item 13HV Desc.: Manhole guards					80433.01					
Vend. Loc.: Woolard Trk Trl			Product code		Unit of measure E		Total cost Total resale value			
3	0	0	0	0	867.00	0.00	867.00	100.00	2,601.00	0 0
Item 13T Desc.: Ridged 300 power vise and thread with cutter and reamer							Serial No. 7532653			
Vend. Loc.: Job 94			Product code		Unit of measure		Total cost Total resale value			
1	0	0	0	0	2,340.00	0.00	2,340.00	100.00	2,340.00	0 0
Item 14 Desc.:					ID No. DW1420-145438		Ditch witch			
Vend. Loc.:			Product code		Unit of measure E		Total cost Total resale value			
1	0	0	0	0	8,632.00	0.00	8,632.00	100.00	8,632.00	0 0

NOTE: MTD = month to date, YTD = year to date, PU = pickup.
*This is a multiple-use form as per individual needs. Since these items are not running stock bought for resale, the "Sell" and "Profit" (which represents sales minus cost) columns are immaterial. The form is primarily for equipment tracking and cost value.

TABLE 9.10 Daily Installation Control Sheet

Job No.__ Date_____ Completion date____ DAILY INSTALLATION CONTROL SHEET FOR COMPLETE

Job Name_____ REPORTING AND FOREMAN EVALUATION

Contract No._____ This sheet must be turned in to office daily

Foreman Signature_____ by 4:30 p.m.

Item Name and No.	Total quantities to be installed	Quantities that must be installed daily	Actual in- stalled today	Total installed to date	Variance	% complete

TABLE 9.11

INVOICE NO: _____
JOB NO: _____
PAGE _____

ELECTRICAL CONTRACTOR'S OR SUBCONTRACTOR'S
APPLICATION FOR PAYMENT

MONTHLY ESTIMATE FOR VOUCHER NO. _____

Contract No. _____   Title _____   Location _____

Subcontractor _____   Subcontractor's address _____

Officer in charge of construction or inspector _____   Estimate for month ending _____

FROM SCHEDULE OF PRICES			TOTAL QUANTITIES			UNIT PRICE				
No.	Item	Unit	Schedule quantity	Previously reported	For month	To date	Delivery	Placing	Total amount this month	Total amount due to date	Remarks

1. Total amount due on basic contract
2. Amount due on change orders
3. Gross amount due to date
4. Reservation 10%
5. Total amount due to date
6. Less previous payments

7. Net amount due to date
8. Less unpaid balance on previous billing
9. Net amount due this request
10. Plus unpaid balance on previous billing
11. Total amount due on contract to date

If check for items has been forwarded please disregard items 8 and 10 and pay only item 9

If a check for less than item 11 has been forwarded deduct the amount of the check from item 11 and remit the balance.

TABLE 9.12

ELECTRICAL CONTRACTOR'S OR SUBCONTRACTOR'S
APPLICATION FOR PAYMENT
MONTHLY ESTIMATE FOR VOUCHER NO. _____

Contract No. *Number*	Title	Location	*Job location*
Subcontractor *Your name*		*Subcontractor's address*	*Your address*
Officer in charge of construction or inspector		Estimate for month ending	*Month, Year*

		FROM SCHEDULE OF PRICES		TOTAL QUANTITIES			UNIT PRICE				
No.	Item	Unit	Schedule quantity	Previously reported	For month	To date	Delivery	Placing	Total amount this month	Total amount due to date	Remarks
1	Demolition (lab)		L.S.	-0-				500.00	500.00	500.00	500.00
2	Concrete (mat)	c.y.	25	-0-	5		50.00		250.00	250.00	250.00
	(lab)	c.y.	25	-0-		5		50.00	250.00	250.00	250.00
3	Framing lumber (mat)	NEBM	3,200	-0-	500		.275		138.00	138.00	138.00
	(lab)	NEBM	3,200	-0-		500		.275	138.00	138.00	138.00
4	Painting (mat)	S.E.	10,000	-0-	2,000		.07		140.00	140.00	140.00
	(lab)	S.E.	10,000	-0-		1,000		.28	560.00	560.00	560.00
5	Clean-up (lab)		L.S.	-0-				250.00	250.00	250.00	250.00

1. Total amount due on basic contract	$2,226.00	
2. Amount due on change orders	-0-	
3. Gross amount due to date	2,226.00	
4. Reservation 10%	222.60	
5. Total amount due to date	2,003.40	
6. Less previous payments	-0-	
7. Net amount due to date		$2,003.40
8. Less unpaid bablance on previous billing		-0-
9. Net amount due this request		2,003.40
10. Plus unpaid balance on previous billing		-0-
11. Total amount due on contract to date		2,003.40

If check for items has been forwarded please disregard items 8 and 10 and pay only item 9

If a check for less than item 11 has been forwarded deduct the amount of the check from item 11 and remit the balance.

Complete" in Table 9.10 is determined by on-site inspection, personal experience, study of the job, and accounting records.

It is important to include all expenses that belong in overhead—and all that belong in direct job expense. It is also important that costs be charged to the proper category; otherwise, control ratios will be meaningless.

It is equally important to remember that control is accomplished by people, and that records reveal the effectiveness or lack of effectiveness of the control. It is on the basis of records that we make decisions for action now and in the future. It is extremely important that cost records be as accurate as possible. Field supervisors must report to management on the labor-hours expended on each category of installation that is set up for detail control. These hours are to be recorded by field supervisors daily by category and submitted to management weekly for compilation. In addition to hours applied, a percentage-complete figure is supplied by category and in total. It is important that all personnel understand the procedure and that these percentages be as accurate as possible, in order that they may be meaningful. Foremen tend to overstate productivity; this should be avoided. Field supervisors should be constantly aware of this tendency so that the *weekly percentage complete column will be considered not only at the end of the week—but daily.*

In contracting, it is through your budget that most records and reports are prepared and routed for proper action, recording, and distribution; therefore, it is extremely important that this department be staffed adequately and capably. It is also important that backup personnel be trained and present or available if possible.

When your operation gets to the P and L statement, it's history. This statement shows you what happened, and it's too late by the time you get it; you can't make corrections backwards.

Another important point to remember in relation to the P and L statement is that the "percentage complete" is a must. Without it, your ratios do not have the needed meaning.

Job files should be stored in the accounting department and kept in their trust. It is extremely important that supplier invoices be checked with purchase-order prices for accuracy. Purchasing should be centralized as much as possible.

It is recommended that the format of the P and L statement and the balance sheet be detailed and that all accounts be included. This includes inventory and retainage. (For inventory, see Table 9.9.)

Summary and General Comments

It is very important that you set your financial house in order—which means including all assets, so that your financial picture will be assessable.

Remember that changes in any system cannot be accomplished without problems. Properly designed changes are the mark of progress. Prospective changes should be first studied and, if felt beneficial, then adopted. The measure of the value of change, of course, is results.

Financial decision making by management does not come easy. Many difficult decisions must be made—and if they are not made, the lack of decision can impair personnel relations. It is important to review all phases of operation before making decisions and then to make them on an organized basis.

Management and Administration

Generally, every business needs some sort of an organization chart (see Figure 3.1.) and an accompanying manual. The purpose of the chart is to graphically depict the various departments, key positions, and lines of authority, as well as the duties, responsibilities, and authority of each key position within the organization. Together, the chart and the manual serve to clarify communications within the company so that all employees will know their job descriptions, what is expected of them, and what the company policy is on all important matters.

It is recommended that you review your past operation, your present situation, and your plan for the future annually, as well as at other important points in your company's history—before setting up the company, before expansion, and after major achievements, for example. The first step is to set an objective for your company. What volume do you want to attain? What types of contracts do you want to accept? In what geographical area do you want to operate? What size contracts do you want to specialize in ? And what is the largest single contract you feel capable of handling? Remember that negotiated contracts are usually more profitable than contracts you have to bid for, because in most instances you must be the lowest bidder in order to receive a contract. *The necessity for bidding low can be very dangerous.*

Once you have set your objectives, put them in writing so that you can refer to them and will not have to rely on memory. From this point on, you can plan your course of action and follow the progress. I suggest that you set up an annual budget and follow it on a monthly basis.

One of the most important areas for management is the existence of backup for *each key position.* This is important.

It is also important that you plan now for the future, and that you lay the groundwork now for future activities in such a manner that you will be set to implement and be guided by your plan. See Table 9.1 for general financial requirements.

Staff meetings should be scheduled whenever necessary, which means whenever you have an agenda sufficient to warrant a meeting. Staff meetings keep key personnel posted on current problems and help to clarify communications. But it is better not to have a staff meeting than to call one with no particular reason.

Production staff meetings should be held before the start of each large job or project, to discuss the entire job. These essential meetings should become standard procedure. However, confidential information need not be discussed with field personnel. It is recommended that you talk in terms of *labor units* or *worker-hours—not dollars—*to field staff. Follow-up is necessary, and progress meetings should be scheduled if needed.

The management-organization-administration aspect of any company is important when considering future plans. Assurance of the profitable and successful perpetuation of your corporation is a factor. Far too often, electrical contractors overlook or play down this area of their operation, considering it necessary only for very large contractors.

Your ability to make a profit depends upon good management, in large part. Management control, planning, and teamwork are essential in any operation. When I mention "planning" and "controls," I am not referring to elaborate procedures or to so-called systems, but to simple but definite planning.

Each key employee must spend his or her time in the most productive ways. Responsibilities should be assigned to and decisions should be made at the lowest competent level.

It is very important to place and consider management aids in proper prospective. The first, prime, and most immediate concern of any organization is making a *net profit*. More sophisticated areas of management training and education are, of course, important—but if these "higher academic" concerns were placed ahead of the fundamental purpose of the profit-making organization, a company could become enmeshed in confusion and frustration and miss the real point of being in business.

Company Policy

The operating manual should contain a section relating to company policy, dealing with the following areas of concern:

1. Purpose of the corporation

2. Corporate officer setup

3. Department setup, divisional setup, etc.

4. Workday and working hours; policy on coffee breaks, sick time, holidays, vacation, etc.; and how time is reported

5. Employee benefits

6. Statement about company loyalty

7. Staff meetings—production and administrative

8. Procedure for handling suggestions for improvement of corporation, etc.

9. Procedures in case of emergency

10. Statement that foremen and general foremen are part of the production management team

11. Statement about personal telephone calls during the workday

12. Procedure for handling expense accounts

13. Any other necessary statements relating to company policy

The primary reasons for establishing clear-cut lines of authority and responsibility are (1) to facilitate smooth and efficient operation in an organization, (2) to give employees at all levels a better opportunity to perform effectively the jobs assigned to them, and (3) to provide a method for assignment or, when necessary, reassignment of duties, and for delegation of responsibility.

The organization chart coordinates the activity of the various departments so that teamwork can be achieved through staff and line activities. To facilitate assignment of job duties and responsibilities, job descriptions are prepared for all positions or job assignments. These job descriptions outline the specific tasks performed in each classification, the qualifications and training necessary to perform such tasks, and the responsibility of the assignment. Some examples of job descriptions are given below.

Job Descriptions

The job descriptions shown below are in accord with general industrial practices and are defined in somewhat broad terms, supplemented when necessary by specific job requirements as mandated by established company policy and procedures.

The job descriptions will provide each employee with an overall outline of job duties and will indicate the qualifications necessary to perform such duties. The job descriptions will also provide managers and department heads with information needed when planning training schedules, other than the regularly established apprenticeship system.

General Superintendent

Authority and Responsibilities. The general superintendent is in charge of the job-management function of Sample Electric Company. Within the limits of company policy, this person is responsible for and has adequate authority to accomplish the duties listed below.

The general superintendent is responsible to the company president for successful accomplishment of duties, and may delegate part of the responsibilities and authority of the position to employees under his or her supervision. The general superintendent is, however, responsible and accountable for the overall results of the job-management function.

In the absence of the president, the general superintendent will assume responsibility for all production activities and will coordinate efforts, when necessary, with the administrative coordinator.

Duties. The general superintendent will:

1. Train foremen in the proper handling of work instructions, including work orders, change orders, drawings, and specifications.
2. Establish procedures for the procuring and scheduling of labor, material, tools, and equipment needed at job sites.
3. Forward all direct job expenses, delivery receipts, and office copies of material requisitions to the shop, no less frequently than weekly.
4. Follow the established company procedure on preparation and review of time cards.
5. Review and utilize the weekly direct labor and monthly job-cost summary reports in supervising jobs.
6. Assist in preparing the monthly estimate for job progress payment requests.
7. Adhere to company procedures on safety rules and working conditions on all jobs.
8. Correlate and supervise the activities of all subcontractors under his or her direction.
9. Correlate the installation of our company's portion of the job with that of the general contractor and other specialty contractors.
10. Follow established procedures in obtaining all necessary licenses and permits.
11. Perform all inspections, i.e., rough inspection, any intermediate inspections that may be needed, and final inspection.
12. Maintain harmonious and effective relations with the owner or awarding authority, architect, and consulting engineer of every job.

13. Participate in the preparation of job-management training programs.

14. Select, train, promote, and release, when necessary, all employees under his or her supervision.

15. Recommend improvements in the policies and procedures which guide the activities of the position.

16. Attend and actively participate in all key personnel conferences.

17. Coordinate activities with those of other key personnel in the accomplishment of company objectives.

Other duties and responsibilities include:

1. Call a meeting of various members of the organization (foreman, day workers, and other workers) at the start of each project to plan and coordinate their work according to company contract responsibilities.

2. Inspect electrical work to ensure that the quality of work is in accordance with specifications.

3. Prepare, or receive from subordinates, periodic reports on progress and costs, and adjust work schedules as overall responsibilities change.

4. Coordinate the work of the project with the estimator and the purchasing agent in any matters concerning their overall responsibilities.

5. Supervise the following subordinate classifications in compliance with company policies and labor-management agreements:
   a. General foreman
   b. Foreman
   c. Day workers and apprentices
   d. Field clerks

Warehouse Supervisor

Authority and Responsibilities. The warehouse supervisor is responsible for the proper storage of materials and equipment used in electrical construction. His or her immediate superior is the general superintendent.

Duties. The duties of the warehouse supervisor include:

1. Control and supervise issuance of tools and equipment; carefully inspect tools and equipment returned from the job sites for repairs, etc.

2. Maintain property records for material, tools, and equipment.

3. Either repair or instruct others in the repair of tools and equipment.

4. Be responsible for and sign all requisitions for all materials, tools, or equipment being removed from the warehouse or returned for credit.

5. Be responsible for delivery of materials to job sites and to and from suppliers.

6. Perform janitorial services for the warehouse office, which includes dusting floors and furniture each morning.

7. Be responsible for the activities and supervision of all his or her subordinates.

Foreman

Authority and Responsibilities. The foreman works directly under the general superintendent.

Duties. The duties and responsibilities of the foreman will be as follows:

1. Be responsible for day workers and apprentices under his or her direction. Plan and coordinate their work in accordance with company contract responsibilities.

2. Prepare periodic reports, as requested by the general superintendent, on labor units applied to specific jobs.

3. Post all direct job expenses to job-cost records.

4. Check all supplier invoices against purchase orders.

5. Assist the estimator and others involved in maintaining material pricing cards and files on a current basis.

6. Be responsible for material shipped to the job or warehoused on the job.

7. Be responsible for construction of electrical circuits and power units, both on new construction jobs and in maintenance service.

8. Be directly responsible for the quality of the work, which must be neat and correct so as to pass electrical inspections.

9. Be responsible for knowing the code and, before any installations are done, making sure that the work planned will pass inspections.

Office Manager

Responsibilities and Duties. The responsibilities and duties of the office manager will be as follows:

1. Be responsible for the records, functions, and activities of corporate secretary.

2. Be responsible for the insurance program of the the corporation.

3. Be responsible for supervision and direction of clerical employees and their assignments.

4. Maintain time sheets for his or her own work hours and for those of subordinates.

5. Assist in the coordination of work of the office with other departments.

6. Make appropriate recommendations in regard to the hiring and discharging of subordinate employees.

Pricing Clerk

Responsibilities and Duties. The responsibilities and duties of the pricing clerk are as follows:

1. Check all suppliers' invoices for prices and extensions, and record any adjustments necessary.

2. Price all requisitions and material credit slips.

3. Direct the activities of the payroll department.

4. Be responsible for the preparation of a monthly profit and loss statement and balance sheet.

5. Calculate various operating and financial ratios and interpret them to the president or board of directors.

5. Actively maintain the accounts payable and accounts receivable ledgers, and all other ledgers and journals maintained by the corporation.

7. Be responsible for all billings on all service-call contracts.

8. Be responsible for the purchase of all capital assets and office supplies.

9. Prepare any and all reports requested by the president.

Estimator

Authority and Responsibilities. Estimating is important to every electrical contractor. It is at this level that all costs need to be included in the estimate or bid, if recovery is to be made and profit realized. This applies to all types of contracts: negotiated, bid, cost-plus, etc.

It is recommended that all personnel involved in estimating attend workshops run by organizations in the electrical industry, such as the

National Electrical Contractors Association (NECA). Your company's estimates should be standardized in reference to the method of costing, etc.; and the format for estimates, once established, should be maintained. Consistency in estimating is of key importance to the profits and record keeping of any company.

In estimating, all three categories of prime cost should be included: *material, direct labor, and direct job cost.* It is essential that the third category—direct job cost—be included in all estimates. Major categories of installation should be set up in the estimate, and these categories should be controlled in cost analysis. The work of staff members involved in estimating needs frequent review by top management in reference to current setup and future plans, especially for backup of key personnel.

Whenever you hire a new staff estimator, remember that this person must estimate in a manner that is in tune with your cost controls, etc. Also remember that *you must maintain a profitable volume in order to support a full-time specialized estimator.*

Most young or inexperienced contractors just getting started spend their time in the following manner. As they grow, these percentages change.

Starting, %	Task	During growth, %
10	Administration	15
0	Public relations and selling	35
15	Estimating	40
0	Job supervision	10
75	Actual working	0

My recommendation is that you consider the following allocation of your time as a guide. The percentages listed here should be considered a goal to be reached whenever practical.

Refer to a good manual on pretakeoff steps to find out how to begin the process of estimating. Most computerized systems are preprogrammed.

Duties. The duties of the estimator include the following:

1. Coordinate the reading of the NECA prequote sheet (both sides) with specifications.

2. On large jobs, do sectionalized takeoffs, and indicate the section (for example, "northeast wing") on the branch circuit sheet. Count all fixtures and color them yellow; indicate the type or types of fixtures, mounting, number and types of lamps, and voltages.

3. Count all devices, and color them all red or green. Draw a green slant line through switches. Note and count only switch covers with two or more gangs as you go. Computerized systems are dif-

ferent; with them, you need only count the number and types of devices, for the components are built into the count.

4. Start the branch circuit takeoff by coloring all two-wire circuit runs with red pencil; as you color each one, press your tally counter. After completing your coloring of a line between any two points, your financial count times 2 equals the number of electrical metallic tubing (EMT) connectors or bushings. Your final count times 3 (where rigid bushings are involved under the floor) will equal the extra feet of wire needed for stub-ups and conduit footage. Your final count times 2 equals the locknut total.

Note. Make note of any special equipment as you go, coloring it pink and drawing in the symbol indicating voltage and whether National Electrical Manufacturers Association (NEMA) 1, 3, 4, or 5.

5. After all branch circuits have been taken off, prepare telephone, intercom, or any other communications circuits on a sheet similar to the feeder sheet. These circuits may all be combined on the same sheet.

6. Prepare a feeder sheet. Include conduit size; number of conduits; footage; type; trench; bush; connectors; all wire sizes, numbers, footages, and types; service head; pull box; etc.

Note. The above system is strictly for manual takeoff. Computerized systems are different.

7. Prepare a sheet for boxes, plaster rings, and blank covers. These counts will come from device counts and from fixture mountings and counts. Greenfield footage can be prepared on this same sheet. This section should be headed "Greenfield footage, wire, connectors, etc." Greenfield footage is obtained by multiplying all recessed fixtures by 6. All recessed fixtures × 2 = Greenfield connector count.

Note. All sheets are to be calculated, checked by the office manager or chief estimator, and returned to the original estimator for transfer to price sheets. The price sheets will be priced by an accountant or estimator and run through a computer or calculator.

8. Transfer the coded price sheets to a summary sheet. Notice carefully any plus or minus factoring and react accordingly. All totals on price summary sheets must match individual sheets; after these are reconciled, pass them on to the office manager or another estimator for checking and finalizing of the bid.

9. Review the specifications again.

10. Visit the site to observe it and note any unusual conditions.

11. If any items are unclear, call the engineer or architect. On the day of bidding, always check with the Dodge Room (where plans and specifications are left with information for the public, on government jobs) or the owner, or both, for addenda and a list of GCs. Twenty-four hours before the bid is due, verify all fixture counts and prices, and panel prices, with at least three suppliers.

Job Supervision

In order to run a profitable contracting operation, it is absolutely imperative to maintain constant supervision over all large contracts. Without proper job supervision, field production can suffer seriously, and you may not know something has gone wrong until a particular job has gone beyond the point of profitable correction.

Completion of jobs on schedule and at a cost equal to or less than the estimate is the prime responsibility of the people responsible for supervising the job. It is therefore necessary that the superintendent and foreman know how the job was figured and how the labor units were applied in the estimate. Any improvements that can be effected not only increase profit on a job but set a basis for better efficiency in the future. Scheduling material for a job is a responsibility of *supervisory personnel*.

Even though the estimate may be initially accurate, if contracts are not supervised constantly and effectively, serious overages and deficiencies can develop. Inadequate supervision can handicap an otherwise profitable operation.

It is important that efforts of key personnel not overlap, particularly in the area of job supervision. Only one superintendent should be assigned to a job, and that supervisor should be held responsible for the effective and profitable production of the job. If members of top management visit a job site, they should be accompanied by the superintendent if possible; if that is not possible, the superintendent should at least be informed about the visit before it takes place. The main reason for this is to avoid overlapping or usurping the superintendent's authority, thus making it less effective. Another reason is that, if field personnel have the opportunity, they will often make a game out of playing top management against the superintendent, at the expense of management control.

Even if a job has been estimated correctly and reasonably, the profit on the entire job can be jeopardized by poor or inadequate supervision. A job must be completed before profit can be realized on it.

Supervision of Personnel

Supervision and superintendents are an extremely vital link in the management chain; a breakdown at this level will stymie the best efforts of even the most competent top management.

The supervisor—in any organization—is under heavy pressure. The job has two conflicting requirements: managing people and achieving production. (See Tables 9.7 and 9.10.) Success is most often measured by the volume of acceptable-quality production within the estimate. As a result, the supervisor is under pressure to achieve short-range effectiveness—often at the expense of long-range requirements for high-level productivity. Supervisors tend to become production-oriented and may tend to neglect the other aspect of their job: managing the people who do the work.

Described below are some negative possibilities within the job of supervision, followed by a description of the qualities which good supervisors cultivate. If any or all of the negative conditions prevail in your supervisory structure, you can be certain that they represent a liability and are costing you profit, among other considerations. On the other hand, if the positive characteristics described below prevail among your supervisory personnel, your company as a whole will benefit.

Snap-Judgment Selection of Employees. Supervision often goes astray at the beginning. Poor selection of an employee could mean years of unhappiness and conflict with fellow employees or associates. *The supervisor or owner who does a poor job of selecting or sizing up staff is inviting trouble.*

Letting the Job Grow Like Topsy. Careless supervisors plus particularly ambitious or lazy employees can shape jobs carelessly. Careless supervisors may assign new duties to any employee who has the capability or the time to squeeze the work in. Ambitious employees sometimes gobble up all duties in sight without regard to whether they are wasting their higher-level skills by carrying a goldbrick. Lazy employees tend to shrug off unpleasant, demanding, or boring duties.

Failure to Make Assignments Clear. Vague or ambiguous instructions like "Let's get crackin' on the big job" or "Will somebody please see that this excess inventory is taken care of?" are ineffective and destructive. No employee can do a good job without adequate authority.

Divided responsibility results in misunderstanding, conflict, and low productivity.

Being a Boss Rather than a Leader. Everybody knows the type of supervisor who says, "When I give an order around here, I want it obeyed!" Pressures on workers beget pressures in management. The "easy way" for a supervisor is to "know it all" and brook no interference. It's much "easier" to handle problems if one doesn't have to consider alternative solutions and possible disadvantages. Supervisors of this sort don't command much respect or loyalty. They inspire little of the "can-do" spirit that is needed to handle the really big, tough jobs with a minimum of problems.

Indifference to Discipline and Recognition. Nothing makes employees more indifferent to discipline and achievement than a supervisor who "couldn't care less." Lax discipline and neglect of job requirements on the part of the supervisor lead to poor work and low productivity.

Too Busy to Train. The supervisor who is too busy getting out production to take time to train employees adequately isn't doing a good job. This kind of supervisor is the person who can never be away from his or her own job.

Playing Everything Close to the Vest. Perhaps worst of all are the supervisors who "keep it all to themselves." They neglect to pass the word. Nobody knows where they stand. Instructions from these people are short and incomplete. These supervisors frown on or reject questions. They typically keep their bosses in the dark—too. The result of this kind of supervision is that turnovers, overloads, slowdowns, and other problems occur unexpectedly.

Qualities of Good Supervisors. Good supervisors, in contrast to those described above, cultivate the ability to deal effectively with the requirements of both production and people. Some of the qualities which they strive to develop are listed below.

1. They choose personnel with care, insight, and discretion. They are aware that proper personnel selection is an important key to solving some of the mysteries of work processes.
2. They practice good job engineering and good planning.

3. They make specific, detailed assignments and then give subordinates the authority needed to accomplish them.
4. In order to engender high morale and high productivity, they demand good-quality work and recognize achievement.
5. When discipline is called for, they administer it promptly and fairly, taking the circumstances into account.
6. They make appropriate and frequent use of praise, letters of appreciation, certificates, and cash rewards—some of the ways in which achievement can be rewarded. They are aware of the importance of *giving recognition* and *openly acknowledging significant achievements*.
7. They communicate well with others, and work constantly to improve their communication skills.
8. They appreciate the value of certain personal attributes, which can be summed up in two pieces of advice:
   a. Be yourself.
   b. Cultivate positive thinking.

Warehouse and Service Department

Recommended procedure is to take an inventory and include it in your balance sheet. Inventory should be adequate but not excessive, so as to avoid overstating assets by purposely overstating inventory.

No material should be removed from the warehouse unless a material requisition is made up and signed for by the person receiving it.

An expense card for each truck should be kept up to date at all times. Whenever a truck gets gasoline from your pumps, the mileage must be recorded on the card; if gasoline is purchased elsewhere, the mileage must be recorded both on the sales slip and on the card.

Scheduling of material and men to jobs is the responsibility of supervisors. Keep in mind that frequent shuttle runs and frequent trips by field personnel are costly. Foremen or others in charge of a job should report, daily and during a specified time range, to the warehouse or supply-house supervisor with requests for material needed for the next day. Following this daily reporting procedure permits scheduling of one delivery a day to each job; any other deliveries should be made on an emergency basis only. Foremen and others involved should be informed of this required procedure.

It is suggested that electrical construction administrators spend their time as follows:

Time, %	Task
20	Administration
35	Public relations and selling
30	Estimating
15	Job supervision

You must maintain tight control over estimating and supervision to assure profitable completion of each job.

Standard Procedure for Precontract Signing and for Administration after Signing

1. Notify your bonding company of the possibility of getting the job.

2. Check out the work reputation and paying habits of the general contractor. This can usually be done by having your supplier use a *Dun and Bradstreet Rating Book*; if this is not possible, check through the local merchants' association.

3. Upon receiving the contract, have your company attorney or a competent company official review it for out-of-the-ordinary legal requirements.

4. If there are one or more subcontracts under the electrical section, such as fire alarm, paging system, or television, follow the procedures in items 1 to 3 above, to check out the firms you plan to subcontract these specialities to. If the firms check out okay, fill out properly the standard contract used by your firm and have the responsible officers of the subcontractor firms sign it. Request either a 100 percent or a 50 percent payment or performance bond from them, whichever the specifications call for. Also request a certificate of insurance, and make sure that you send the firms affected one set of plans and specifications.

5. When item 1 to 3 have been sucessfully accomplished, request a construction schedule from the GC.

6. Set up a submittal control form for shop drawings, and push the suppliers *as hard as you have to*, until they give you eight sets of shop drawings.

7. The superintendent, foreman, and manager or estimator of your company must attend the GC's preconstruction conference. They should all be prepared to take notes and ask questions.

8. Hold a preconstruction conference within your own company. Employees that should attend are the superintendent, foreman, helper, estimator, and office manager. All items brought up at the

main GC's preconstruction conference must be gone over thoroughly. For other items which should be discussed at the preconstruction conference, refer to the discussion above.

9. The superintendent is aware of his or her responsibility for the movement of work force, machinery, materials, etc., but must check with the purchasing department to find out where to purchase these materials most economically, and also to obtain a tentative shipping schedule. This is the procedure to be followed for miscellaneous material pickups. The superintendant does not, of course, personally make actual purchases, pickups, and deliveries, which are usually acomplished by the material runner.

10. Set up the job shack (storage and office trailers) movement, and make a temporary setting of the power pole.

11. A construction work-force schedule must be made out by the estimator, using a bar-graph chart or a progress schedule.

12. All activities involving electrical jobs or specialty subcontractors under contract to your company must be kept under supervision.

13. No one shall make any changes or do any extra work without written consent obtained through the GC's superintendent and our office. Only a worker's immediate supervisor can authorize him or her to do any extra work, or any work not shown on plans or specifications. Exceptions to this rule are allowed only in situations involving a threat to life, limb, or property.

APPENDIX

Administrative Package for Use in Dealing with Subcontractors

Included in this section are many documents—form letters, invoices, applications for payment, etc.—which will be of use in dealing with subcontractors. See Figures 9.1 to 9.13.

ATTENTION:_____   Date_____

To Whom it May Concern:
    This letter is to inform you of the cost exposure we may be subjected to by use of owner-furnished fixtures, devices, equipment, etc. If we should incur such costs, because of any of the circumstances listed below, we have no choice but to advise you and then pass this cost on to the customer, owner, or G.C.

1. Material delivery and scheduling delays

2. Warranty problems

3. Malfunctioning materials

4. Concealed damages

5. Excess materials and detail work

6. Late shop drawings

    We hope that none of the above circumstances will arise. It is, however, our duty to advise you.

                                                                    Respectfully,

Figure 9.1   Hazards of owner-furnished materials.

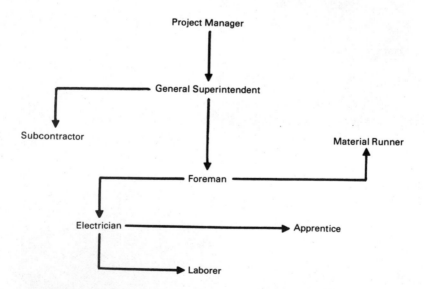

Figure 9.2   Functional organization chart.

CONTRACT NO._____

RE:_____

DATE_____

To Whom It May Concern:

It is indeed a pleasure to have this opportunity to work with your firm on the referenced contract. We sincerely hope to have a very smooth transition from start to completion with minimal interruptions during the performance of your segment of work. To accomplish this end, it is of primary importance that you *expedite* your *submittals* in a timely fashion and in as complete a form as the contract specifications require.

Oftentimes specifications will give a listing of several material items which have differing physical characteristics but similar intended uses. In this case, one or more or possibly all of these items may be used to provide a complete system. In some instances, it may be construed as the contractor's option to use the item or items of his or her choice. To help eliminate errors on our submittal control form and to avoid unnecessary submittal requirements on your behalf, we request that you forward to us a complete listing of the material items you intend to submit for review and approval.

This is a contractor's quality-control contract; therefore, each and every item, shop drawing, etc., must bear your signed approval before you submit the data to this office. The following are additional requirements you must satisfy for all submittal work: You shall submit the data as soon as possible; reference the data per the "specific" specification section, paragraph, subparagraph, or both; submit ample numbers of copies as required; when the data is not or cannot be referred to by a particular federal or commercial standard as specified, the data must be accompanied by certifications from either the manufacturer or your firm attesting that the materials comply with the referenced federal or commercial standards; when data is referenced by a commercial standard and the specification requires that the data must meet a federal standard, you must submit, as support, a copy of that commercial standard along with the subject data.

In the event that you deviate from the specifications, your data shall clearly denote "This is a deviation" and the data must be accompanied by a certification. You shall indicate the lead times required for delivery of all materials.

Attached are sample formats for your use in preparing submittal correspondence to this office. We urge you to comply with our aforementioned requests, as failure to do so may result in disapproval of data or more severely, the rescinding of your subcontract.

Respectfully,

Enclosure.

Figure 9.3 Letter to a subcontractor.

TO _____   DATE _____

_____

_____

PROJECT _____

PROJECT NO. _____

LOCATION _____

To Whom It May Concern:
    In review of our records for the above-captioned project, please furnish to our office the information indicated below.
    No payments will be made on the contract until all information is furnished and in order.

( )   Identifying number or social security number.
( )   Certificate of worker's compensation, general liability insurance, and automobile liability insurance in accordance with the applicable state law and the contract documents.
( )   Please send us on your letterhead a list of your key personnel who have knowledge of the above-referenced project. Names, addresses and telephone numbers are required.
( )   Thirty-day notice of cancellation to general contractor.
( )   Renewal certificate not received by this office. Please comply immediately.
( )   Cancellation notice has been received. Please furnish us with reinstatement notice or replacement certificate.
( )   Submit payroll forms weekly (social security number and address for each employee must be listed the first time the name appears on the payroll). *Please submit 3 copies of each weekly payroll report every seven (7) days.* Submit payroll only directly to _____ .
( )   Please send us a list of names, social security numbers, height, and weight for all employees whom you anticipate will be working on this job. Badges must be made up for each of these employees before they can start work on the job; therefore, we must have this information as soon as possible for federal projects.
( )   Monthly vouchers will only be accepted on our subcontractors' applications-for-payment forms.

Figure 9.4   Project administration.

Date_____

RE: Payroll Reports

CONTRACT NO. _____

To Whom It May Concern:
    It is very important that you submit your payroll reports to us in a timely fashion, every seven (7) days. Failure to submit these payrolls in a timely fashion may result in withholding of your funds.

Sincerely yours,

Figure 9.5   Payroll reports.

Date _____

TO: _____

_____

_____

RE: _____

CONTRACT NO. _____

SUBJECT: Submittal Checklist

To Whom It May Concern:

   Please be advised that *all* materials warrant an approval before delivery on the project. Therefore, the expedition of submittals is of great importance and must be forwarded immediately after receipt of your contract. To avoid delays on payment and approval, certain criteria should be applied to each submittal *before* submitting to our office. They are as follows:

_____ 1. Reference all submittal information to the specifications.

_____ 2. Highlight each specific item or items being submitted for approval.

_____ 3. *Do not* deviate from the specifications.

_____ 4. Submit sufficient copies (8).

_____ 5. Add your company's stamp of approval, signed and dated.

_____ 6. All certifications must conform to the project's name, location, and contract number and specifications.

_____ 7. *On* each submittal note the specification section, paragraph, and subparagraph (if needed).

   We would appreciate your full cooperation in this matter to ensure smooth operation on this project.

Thank you,

Figure 9.6   Submittal checklist.

STATEMENT OF COMPLIANCE,
FOR GOVERNMENT CONTRACTORS

Payroll Number	Payroll Payment Date	Contract Number

Date_____

I, _____, _____ do hereby state:
　　　Name of Signatory Party　　　　　　　　Title

(1) That I pay or supervise the payment of persons employed by _____
　　　　　　　　　　　　　　　　　　　　　　　　　　　Contractor or Subcontractor
on the _____ , that during the payroll commencing on the_____
　　　　　Building or Work
day of _____, 19___, and ending on the _____ day of _____ , 19___ ,
all persons employed on said project have been paid the full weekly wages earned, that
no rebates have been or will be made either directly or indirectly to or on behalf of
said _____ from the full weekly wages earned by any person,
　　　Contractor or Subcontractor
and that no deductions have been made either directly or indirectly from the full wages earned by
any person, other than permissible deductions as defined in Regulations, Part 3 (29CFR Subtitle
A), issued by the Secretary of Labor under the Copeland Act, as amended (48 Stat. 948.63 Stat.
108, 72 Stat. 967; 76 Stat. 357; 40 U.S.C. 276c), and described below.

_____
_____
_____
_____

(2) That any payrolls otherwise under this contract required to be submitted for the above period are
correct and complete; that the wage rates for laborers or mechanics contained therein are not
less than the applicable wage rates contained in any wage determination incorporated into the
contract; that the classifications set forth therein for each laborer or mechanic conform with the
work performed.

(3) That any apprentices employed in the above period are duly registered in a bona fide O.S.C.
program registered with a state apprenticeship agency recognized by the Bureau of Apprenticeship and Training, United States Department of Labor.

(4) That

(a) WHERE FRINGE BENEFITS ARE PAID TO APPROVED PLANS, FUNDS, OR PROGRAMS: In addition to the
☐ basic hourly wage rates paid to each laborer or mechanic listed in the above-referenced
payroll, payments of fringe benefits as listed in the contract have been or will be made to
appropriate programs for the benefit of such employees, except as noted in section 4c
below.

(b) WHERE FRINGE BENEFITS ARE PAID IN CASH: Each laborer or mechanic listed in the above-
☐ referenced payroll has been paid as indicated on the payroll, an amount not less than the
sum of the applicable basic hourly wage rate plus the amount of the required fringe benefits as listed in the contract, except as noted in section 4c below.

(c) EXCEPTIONS:

Exception (Craft)	Explanation

REMARKS

Name and Title	Signature

The willful falsification of any of the above statements may subject the contractor or subcontractor to
civil or criminal prosecution. See Section 1001 of Title 16 and Section 231 of Title 31 of the United
States Code.

DD FORM 879　　Previous Editions are Obsolete.　　s/n-0102-lf-008-6901

Figure 9.7 Statement of compliance.

STATEMENT OF ACKNOWLEDGMENT, FOR GOVERNMENT CONTRACTORS

Name and Address of Prime Contractor
(Include Zip Code)

Name and Address of Subcontractor
(Include Zip Code)

Prime Contract Number   Date Subcontract awarded

The prime contractor whose signature appears below states in accordance with the provisions of the clause entitled "Subcontractors" of the above-numbered contract with the UNITED STATES OF AMERICA that a subcontract was awarded on the date shown above by_____
to the subcontractor identified above to perform the following work:

Subcontract Amount   Date Signed (Day, Month, Year)

☐ $50,000 or over

☐   Under $50,000

Project

Typed Name and Title of Persons
Signing for Contract

Location   By (Signature)

The subcontractor whose signature appears below acknowledges accordance with the provisions of the clause entitled "Subcontractors" contained in contract No._____
referred to above, that the following provisions of the prime contract are incorporated into and made a part of the subcontract:

Equal opportunity
Contract Work Hours Standard
    Act—Overtime Compensation
Payrolls and payroll records
Withholding of funds

Davis Bacon Act
Apprentices
Compliance with Copeland Regulations
Subcontracts
Contract termination—debarment

Date Signed
(Day, Month, Year)

Typed Name and Title of Person Signing
for Subcontractor

By (Signature)

DD Form 1566   SM0102-LF-014-8000   U.S. Government Printing Office 1973

Figure 9.8   Statement of acknowledgment.

REPRESENTATIONS AND CERTIFICATIONS, FOR GOVERNMENT CONTRACTORS
(Contractors and Architect-Engineer Contract)
(For use with Standard Form 19.21 and 252)

Name and Address of Bidder   Date of Bid
(No., Street, City, State, and Zip Code)

In negotiated procurements,"bid" and "bidder" shall be construed to mean "offer" and "offerer." The bidder makes the following representations and certifications as part of the bid identified above. (Check appropriate boxes.)

1. SMALL BUSINESS
    He or she ☐ is, ☐ is not, a small business concern. A "small business concern" for the purpose of government procurement is a concern, including its affiliates, which is independently owned and operated, is not dominant in the field of operations in which it is bidding on government contracts, and can further qualify under the criteria concerning number of employees, average annual receipts, or other criteria as prescribed by the Small Business Administration. For additional information, see governing regulations of the Small Business Administration (13 CFR Part 121).

2. MINORITY BUSINESS ENTERPRISE
    He or she ☐ is, ☐ is not, a minority business enterprise. A "minority business enterprise" is defined as a "business, at least 50 percent of which is owned by minority group members or, in case of publicly owned businesses, at least 51 percent of the stock of which is owned by minority group members." For the purpose of this definition, minority group members are Negroes, Spanish-speaking American persons, American-Orientals, American-Indians, American-Eskimos, and American-Aleuts.

3. CONTINGENT FEE
    (a) He or she ☐ has, ☐ has not, employed or retained any company or person (other than a full-time bona fide employee working solely for the bidder) to solicit or secure this contract, and (b) he or she ☐ has, ☐ has not, paid or agreed to pay any company or person (other than a full-time bona fide employee working solely for the bidder) any fee, commission, percentage, or brokerage (e.g., contingent upon or resulting from the award of this contract) and agrees to furnish information relating to items a and b above as requested by the contracting officer. (For interpretation of the representation, including the term "bona fide employee," see Code of Federal Regulations, Title 41, Subpart 1-1.5.)

4. TYPE OF ORGANIZATION
    He or she operates as an ☐ individual, ☐ partnership, ☐ joint venture, ☐ corporation, incorporated in the state of_____ .

5. INDEPENDENT PRICE DETERMINATION
    (a) By submission of this bid, each bidder certifies, and in the case of a joint bid, each party thereto certifies as to his or her own organization, that in connection with this procurement:
        (1) The prices in this bid have been arrived at independently, without consultation, communication, or agreement for the purpose of restricting competition, as to any matter relating to such prices with any other bidder or with any competitor;
        (2) Unless otherwise required by law, the prices which have been quoted in this bid have not been knowingly disclosed by the bidder and will not knowingly be disclosed by the bidder prior to opening, in the case of a bid, or before award, in the case of a proposal, directly or indirectly to any other bidder or to any competitor; and

Figure 9.9 Representation and certifications.

(3) No attempt has been made or will be made by the bidder to induce any other person or firm to submit or not to submit a bid for the purpose of restricting competition.

(b) Each person signing this bid certifies that:

(1) He or she is the person in the bidder's organization responsible within that organization for the decision about the prices being bid herein, and he or she has not participated, and will not participate, in any action contrary to *a*1 through *a*3 above; or

(2) (i) He or she is not the person in the bidder's organization responsible within that organization for the decision about the prices being bid herein, but has been authorized in writing to act as agent for the persons responsible for such decision in certifying that such persons have not participated, and will not participate, in any action contrary to *a*1 through *a*3 above, and as their agent does hereby so certify; and (ii) he or she has not participated, and will not participate, in any action contrary to *a*1 through *a*3 above.

(c) This certification is not applicable to a foreign bidder submitting for a contract which requires performance or delivery outside the United States, its possessions, and Puerto Rico.

(d) A bid will not be considered for award where: *a*1 through *a*3 or *b*, above, has been deleted or modified. Where *a*2, above, has been deleted or modified, the bid will not be considered for award unless the bidder furnishes with the bid a signed statement which sets forth in detail the circumstances of the disclosure and the head of the agency, or designee, determines that such disclosure was not made for the purpose of restricting competition.

NOTE: Bids must set forth full, accurate, and complete information as required by this invitation for bids (including attachments). The penalty for making false statements in bids is prescribed in 18 U.S.C. 1001.

19.304	(REV. 1980 AUG)	Standard Form 19-8, June 1976 Edition, General Services Administration, Fed. Proc. Reg. (41 CFR) 1-16 401 and 1-16 701

Appendix A is hereby made a part hereof.

APPENDIX A TO STANDARD FORM 19-8

WOMAN-OWNED BUSINESS (SEPTEMBER 1978)

The offeror represents that the firm submitting this offer ☐ is, ☐ is not, a woman-owned business. A woman-owned business is a business which is at least 51 percent owned, controlled, and operated by a woman or women. "Controlled" is defined as exercising the power to make policy decisions. "Operated" is defined as actively involved in day-to-day management. For the purposes of this definition, businesses which are publicly owned, joint stock associations, and business trusts are exempted. Exempted businesses may voluntarily represent that they are or are not women-owned, if this information is available.

PERCENTAGE OF FOREIGN CONTENT (SEPTEMBER 1978)

Approximately _____ percent of the proprosed contract price represents foreign content or effort.

SMALL DISADVANTAGED BUSINESS CONCERN (AUGUST 1980)

(a) The offeror represents that he or she ☐ is, ☐ is not, a small business concern owned and controlled by socially and economically disadvantaged individuals. The term "small business concern" means a small business as defined pursuant to Section 3 of the Small Business Act and relevant regulations promulgated pursuant thereto. The term "small business concern

Figure 9.9 *(Continued)*

owned and controlled by socially and economically disadvantaged individuals" means a small business concern

(1) That is at least 51 percent owned by one or more socially and economically disadvantaged individuals; or, in the case of any publicly owned business, at least 51 percent of the stock of which is owned by one or more socially and economically disadvantaged individuals

(2) Whose management and daily business operations are controlled by one or more such individuals

(b) The offeror shall presume that socially and economically disadvantaged individuals include black Americans, Hispanic Americans, native Americans (i.e., American Indians, Eskimos, Aleuts, and native Hawaiians), Asian-Pacific Americans (i.e., U.S. citizens whose origins are in Japan, China, the Philippines, Vietnam, Korea, Samoa, Guam, the U.S. Trust Territories of the Pacific, Northern Marianas, Laos, Cambodia, and Taiwan), and other minorities or any other individuals found to be disadvantaged by the Small Business Administration pursuant to Section 8A of the Small Business Act.

EQUAL EMPLOYMENT COMPLIANCE (SEPTEMBER 1978)
By submission of this offer, the offeror represents that, to the best of his or her knowledge and belief, except as noted below, up to the date of this offer no written notice (such as a show-cause letter, a letter indicating probable cause, or any other written notification citing specific deficiencies) has been received by the offeror from any federal government agency or representative thereof indicating that the offeror or any of its divisions or affiliates or known first-tier subcontractors is in violation of any of the provisions of Executive Order 11246 of September 24, 1965, as amended, or rules and regulations of the Secretary of Labor (41 CFR, Chapter 60) and specifically as to having an acceptable affirmation action compliance program or being in noncompliance with any other aspect of the Equal Employment Opportunity Program.

Figure 9.9 *(Continued)*

QUALITY CONTROL AND SAFETY, FOR GOVERNMENT CONTRACTORS

This quality-control system includes the three phases of control management.

A. PREPARATORY INSPECTION. Preparatory inspection shall be performed before beginning any work, and in addition, before beginning each segment of work. Preparatory inspection shall include a review of the contract requirements, a review of the contract for shop drawing and other submittal data, a check to assure that required control testing will be provided, a physical examination to assure that all materials and equipment conform to approved shop drawings and submittal data, and a check to assure that all required preliminary work has been completed.

B. INITIAL INSPECTION. An initial inspection shall be performed as soon as a representative segment of the particular item of work has been accomplished. Initial inspection shall include performance of scheduled tests, examination of quality of workmanship, a review of test results for compliance with contract requirements, a review of omissions or dimensional errors, and approval or rejection of the initial segment of work.

C. FOLLOW-UP INSPECTIONS. Follow-up inspections shall be performed daily, and more frequently as necessary, and shall include continued testing and examinations to assure compliance with the contract requirements.

The following additional safety measures will be instituted during the performance of the contract:

A. Weekly safety meetings with workers and monthly safety meetings with subcontractors.

B. An approved first-aid kit will be maintained in a designated location in the office trailers.

C. The project superintendent will maintain a first-aid log and will act as first-aid attendant.

D. A list of emergency telephone numbers will be posted in the office trailers. General Hospital has been chosen for use in all emergencies.

E. This saftey program is in compliance with U.S. Army Corps of Engineers Safety and Health Requirements Manual EM-185.

Figure 9.10 Quality control and safety.

PURCHASE-ORDER LANGUAGE, FOR GOVERNMENT CONTRACTORS

1. Executive Order (EO) 11246, Section 202, of September 24, 1965, requires verbatim incorporation of said section into all contracts. During the performance of this contract, the contractor agrees as follows:

   (a) The contractor will not discriminate against any employee or applicant for employment because of race, sex, creed, color, national origin, or age. The contractor will take affirmative action to ensure that applicants are employed and that employees are treated equally during employment, without regard to their race, sex, creed, color, national origin, or age. Such action shall include, but not be limited to, the following: employment, upgrading, demotion, or transfer, recruitment or recruitment advertising, layoff or termination, rate of pay or other forms of compensation, and selection for training including apprenticeship. The contractor agrees to post in conspicuous places, available to employees and applicants for employment, notices to be provided by the contracting officer setting forth the provisions of this nondiscrimination clause.

   (b) The contractor will, in all solicitations or advertisiments for employees placed by or on behalf of the contractor, state that all qualified applicants will receive consideration for employment without regard to race, sex, creed, color, national origin, or age.

   (c) The contractor will send to each labor union or representative of workers with which the company has a collective bargaining agreement or other contract or understanding, a notice to be provided by the agency contracting officer, adviser, the labor union or workers' representative of the contractor's commitments under Section 202 of Executive Order 11246, and shall post copies of the notice in conspicuous places available to employees and applicants for employment.

   (d) The contractor will comply with all provisions of Executive Order 11246, and with the rules, regulations, and relevant orders of the Secretary of Labor.

   (e) The contractor will furnish information and reports required by Executive Order 11246 and by the rules, regulations, and orders of the Secretary of Labor or pursuant thereto, and will permit access to company books, records, and accounts by the contracting agency and the Secretary of Labor for purposes of investigation to ascertain compliance with such rules, regulations, and orders.

   (f) In the event of the contractor's noncompliance with the nondiscrimination clauses of this contract or with any of such rules, regulations, or orders, this contract may be cancelled, terminated, or suspended in whole or in part and the contractor may be declared ineligible for further government contracts in accordance with procedures authorized in Executive Order 11246, and such other sanctions may be imposed and remedies invoked as provided in Executive Order 11246, or by rule, regulation, or order of the Secretary of Labor, or as otherwise provided by law.

   (g) The contractor will include the provisions of paragraphs a to g in every subcontract or purchase order unless exempted by rules, regulations, or orders of the Secretary of Labor issued pursuant to action 204 of Executive Order 11246, so that such provisions will be binding upon each subcontractor or vendor. The contractor will take such action with respect to any subcontract or purchase order as the contracting agency may direct, as a means of enforcing such provisions including sanctions for noncompliance. Provided, however, that in the event the contractor becomes involved in or is threatened with litigation with a subcontractor or vendor as a result of such direction by the contracting agency, the contractor may request the United States to enter into such litigation to protect the interest of the United States.

2. Executive Order 11246, Section 204, allows the Secretary of Labor to exempt certain contracts.

3. 41 CFR 60-1.5a exempts contracts of less than $10,000.

4. 41 CFR 60-1.4d allows incorporation of Section 202 of Executive Order 11246 by reference in contracts not exceeding $50,000. See the purchase-order statement below.

   NONDISCRIMINATION: The vendor will comply with all provisions of Section 202 of Executive Order 11246, as amended by Executive Order 11375 or subsequent executive orders and by the rules and regulations set forth by the Secretary of Labor in effect as of the date of this order.

Figure 9.11 Purchase-order language.

THE CERTIFICATION OF NONSEGREGATED FACILITIES

The (_____Company) assures GOVERNMENT CONTRACTORS and CONCERNED FEDERAL AGENCIES that we do not and will not maintain or provide for our employees any segregated facilities at any of our establishments, and that we do not and will not permit our employees to perform their services at any location, under our control, where segregated facilities are maintained. The company understands that the phrase "segregated facilities" includes facilities which are, in fact, segregated on a basis of race, sex, color, creed, national origin, or age, because of habit, local custom, or otherwise. The company understands and agrees that maintaining or providing segregated facilities for our locations, under our control, where segregated facilities are maintained is a violation of the equal opportunity clause required by Executive Order 11246 of September 24, 1965.

The company further understands and agrees that a breach of the assurance herein contained subjects us to the provisions of the orders of the Secretary of Labor and the provisions of the equal opportunity clause enumerated in contracts or referenced on purchase orders by the government and government contractors.

Figure 9.12 Equal compliance (facilities).

RE: _____

SUBCONTRACTOR QUALITY-CONTROL REQUIREMENTS

A. The subcontractors will be required to designate quality-control project representatives (QCPRs) who shall inspect initially and from time to time the continuing quality of the work in progress. The subcontractors will be held responsible for reviewing their shop drawings and samples before submitting these to the general contractor. Their QCPRs shall relate to the contractor's QCPR any information, comment, and other data deemed necessary to establish and confirm the quality levels of the work in their particular trades, and shall submit evidence in the form of notarized manufacturer's certificates, inspection reports, etc., to establish these quality levels for the work for which they are responsible.

B. Each subcontractor's QCPR shall submit to the project's contractor's QCPR a letter of authority, signed by a corporate officer, to the effect that the subcontractor's QCPR has authority to stop the portion of work his or her firm is involved in and to reinitiate such work when satisfied that the expected quality control has been reached, and that such authority supersedes that of any other on-site employee of any subcontractor.

C. The subcontractors' QCPRs will be responsible to the contractor's QCPR for making sure that all predelivery testing-shop drawings, lists of materials and equipment, catalog cuts, color charts, samples, and certifications which are required by their respective portions of the work are available on the job.

D. The subcontractor's QCPRs shall have the duties and responsibilities of the contractor's QCPR outlined above, under the general supervision of the contractor's QCPR. The subcontractors' QCPRs shall have the duty and the authority to advise the contractor's QCPR as well as their own employees of any poor-quality work discovered.

Finally, we are aware that whoever knowingly and willfully makes any false, fictitious, or fraudulent representation may be liable to criminal prosecution under 18 U.S.C. No. 100.

_____
Signature and Title of Highest Official

Figure 9.13 Subcontractor quality control.

Five-Year Plan

I. *General statements leading up to present planning and future planning.* A deficiency in any of the three elements that have a direct bearing on the value of a contracting firm will affect the company's value. These three elements are
   a. Profits, both historical and current
   b. The quality of the management team
   c. The company's plans, both for the near future and long-term
II. *Present planning.* This usually includes 1 year at a glance, which is no more than a budget for the upcoming year.
III. *Future planning.*
   a. This road map to the company's future usually includes 2 to 5 years into the future. Future planning is a conceptual process that establishes a course of action which has been identified by first setting basic company objectives. It forces owners to verbalize their basic desires intelligently and gives management a blueprint. Contracting firm owners do have or should have expectations for the future of their investment.
   b. Without future planning to direct management, the firm will tend to drift from year to year, and expectations may never be fulfilled.
   c. In small and medium-size companies, the president or executive vice president normally would be the logical choice to lead future planning. This person must be familiar with the various planning techniques and must understand the effort entailed.
IV. *Analysis of objectives.*
   Most companies have never been required or found it necessary to define their corporate objectives. To move forward, the owners, stockholders, or family must clearly define what they want the company to be. (The Example following shows how.)

Example. To undertake $5 million in sales, the company's net worth must be $500,000 minimum. Our goal is to attain this net worth in 5 years. We must obtain a $100,000 minimum increase in net worth per year, and must maintain a minimum of $375,000 cash in the bank.

Sales	$1,232,000
Divided by number of jobs	22
Average size of jobs	$ 56,000
Sales 1987–1988	$ 1,232
Average number of jobs	22
Average size of jobs	$ 56,000
Average number of field employees	30
Average number of OH employees	5
Average sales per field employee	$ 41,067
Average sales per OH employee	$ 246,800
Company	Sample Electric Company
5-year sales goal	$5,000,000
Average number of jobs	20
Average size of jobs	$ 250,000
Average number of field employees	50
Average number of OH employees	8
Average sales per field employee	$ 100,000
Average sales per OH employee	$ 625,000

Note. In the electrical contracting industry, a company's sales should be based on a turnover of 5 to 8 times its net worth.

In the table below, blanks are to be filled in by the reader, as an exercise. Refer to the table above for examples of how to fill in these blanks.

Sales	$_____
Divided by number of jobs	_____
Sales for year	$_____
Average number of jobs	_____
Average size of jobs	_____
Average number of field employees	_____
Average number of OH employees	_____
Average sales per field employee	$_____
Average sales per OH employees	$_____
For _____ -year goal of $ _____ in sales	
Average number of jobs	_____
Average size of jobs	_____
Average number of field employees	_____
Average number of OH employees	_____
Average sales per field employee	$_____
Average sales per OH employee	$_____

To undertake $_____ in sales, the company's net worth must be $_____. Our goal is to attain this net worth in_____ years. We must obtain a $_____ minimum increase in net worth per year, and must maintain a minimum of $_____cash in the bank.

A company's objectives may be:

a. To stay small
b. To increase market share
c. To generate as much profit as possible
d. To eliminate competition
e. To keep family or qualified employees in control of the company
f. To diversify into other market areas
g. To become the geographical industry leader
h. To become the technical industry leader

Note. Establishing any or all of the above is not future planning; strictly speaking it is corporate planning, but no company can formulate a future plan until the corporate objectives have been set.

V. *Future planning.* Presently _____ is a small firm with _____employees during winter and _____ during its peak season in the summer. Its sales are $_____ annually. The future planner or director will be _____. The future planning committee members will be_____

_____.The members of this committee must know very explicitly, beyond a doubt, what is expected of them, and this includes drafting a general plan for the future. The company's divisions are as follows:_____

_____.

Our immediate objectives over the next 2 years (19___–19___) are as follows:

a. To maintain and control our growth. Our growth is not to exceed _____ percent of our previous sales; _____ percent of this will be inflation.
b. To generate as much profit as possible. How?
   (1) By better planning and scheduling
   (2) By increasing the number of distributors we purchase from, thereby obtaining the best price
   (3) By paying bills in a manner to take all discounts
   (4) By maintaining $_____ cash in the bank at all times
   (5) By enhancing our bondability

(6) By increasing cashflow, which means giving an 0.08 percent discount on all federal contracts

VI. *Improving personnel.* During the _____ year, the plan is to improve the grass roots of all departments by staffing them with supercompetent personnel and by providing better training for existing employees. At this point, any one of us who doubts the value of this future planning must ask himself or herself, "What will happen if no action is taken?" Since the only alternative to future planning is corporate drift, even the most doubtful skeptic should be ready to proceed with details of a general plan.

VII. *Critical analysis of the company's strengths and weaknesses.* The planning committee at this point must move to begin a critical analysis of the company since each company has certain strengths and weaknesses. Perhaps a strong marketing department has been able to overcome technical inadequacies. Perhaps lack of competition has been a key element of success. No matter what our strengths may be, identifying them is one of the greatest benefits future planning has to offer.

VIII. *Areas of critical analysis.* The contractor can best benefit by analyzing strengths and weaknesses in seven important areas:

  a. *Ownership.*
    (1) Who will be the future owners of the company?
    (2) How will the perpetuation of the company be guaranteed?
  b. *Sales.* What should sales figures be 5 years from now? Note that all sales projections are the key to all the company's requirements in future years; if volume, for example, will be low, greater attention should be focused on profits. On the other hand, large increases in sales will probably mean sacrifice in other functional areas of low initial returns.
  c. *Market segment.* Will our specialty, whether it is residential, commercial, industrial, highway, or substation, support its future sales objective? If not, steps must be taken to develop new capabilities that will ultimately result in increased profits.
  d. *Geographical.* Does our firm's present location have the potential for continued or increased volume? If not, union agreements, state laws, and work-force mobility should be analyzed to develop a list of geographical areas to which the firm might relocate or expand.
  e. *Field personnel.* Will the company's present field force be able to handle the installations generated by future sales in

geographical areas defined in future planning? Retirements, technological improvements, increased complexity of operations, and training programs have a direct bearing on the answer to this question.

f. *Management.* Is the company's management capable of achieving the results? Managers must possess an understanding of the present as well as the ability to cope with the future. The present management structure should be compared with future goals. Where will the firm's strengths and weaknesses be? Who should be replaced? Who has the potential for assuming more responsibility and therefore should be cultivated?

g. *Financial structure.* Considering the firm's present capital structure and corporate objectives, what demands will be placed on the financial structure of the company? Whether large cash outlays, forsaken profits for sales growth, or bank loans are required, future planning will have a significant impact on capital requirements. Once these areas have been examined, and strengths and weaknesses have been identified, a statement should be combined with the objective of the company owner; this becomes the future planning. Clearly, the basic statement should be written in some detail to facilitate its implementation. This statement will become the road map of our company, giving management the direction it needs to provide for the long-range vitality of our firm, not just our short-term survival.

Finally, a company cannot be successful year after year without future planning. Furthermore, only with future planning can it move in the directions it wants to go, at the pace and profitability level it desires—making decisions based on choices rather than reacting based on situations.

Index

ABOUT THE AUTHOR

Charles L. Ray, Jr., now retired, is a former electrical contractor and business, marketing, and technical consultant. For 25 years, he owned and operated the firm of C. L. Ray, Jr., Inc., Electrical Contractors, which employed 35 people and specialized in residential, commercial, and industrial contracting. During that time he was able to secure several very large government contracts. Mr. Ray has also led many seminars for electrical contractors on management and technical services. He currently resides in Roanoke, Virginia.